Oscar Oldberg

The Metric System in Medicine

containing an account of the metric system of weights and measures, Americanized

and simplified, a comprehensive dose table, and three hundred practical

illustrations of metric prescription writing

Oscar Oldberg

The Metric System in Medicine
containing an account of the metric system of weights and measures, Americanized and simplified, a comprehensive dose table, and three hundred practical illustrations of metric prescription writing

ISBN/EAN: 9783337349301

Printed in Europe, USA, Canada, Australia, Japan

Cover: Foto ©berggeist007 / pixelio.de

More available books at **www.hansebooks.com**

THE

METRIC SYSTEM IN MEDICINE

CONTAINING AN ACCOUNT OF THE

METRIC SYSTEM OF WEIGHTS AND MEASURES, AMERICANIZED AND SIMPLIFIED,

A COMPREHENSIVE

DOSE TABLE,

AND

Three Hundred Practical Illustrations of Metric Prescription Writing, Selected from Recipes in Actual Use in Hospital and Out-Door Practice.

BY

OSCAR OLDBERG, PHAR. D.,

Medical Purveyor, United States Marine Hospital Service; Professor of Materia Medica, National College of Pharmacy, Washington, D. C.; Member of the Sixth Decennial Committee of Revision of the Pharmacopœia of the United States, etc.

PHILADELPHIA:

PRESLEY BLAKISTON,

1012 WALNUT STREET.

1881.

PREFACE.

    AMONG the necessary conditions to the ultimate adoption
of the Metric System in medicine and pharmacy, a
willingness on the part of the younger members of the
professions to take the trouble to use it, and on the part of
the colleges to teach it, are most important. The
recognition of the superiority of this system of weights and
measures, and of the desirability of its adoption, has been
general and earnest, and the new pharmacopœia, the
working formulæ of which will be in parts by weight, will
give a new and powerful impetus to this reform.

    This little volume is offered as a guide to those members
of both professions who desire a special handbook on the
subject, and especially to the students in the colleges. It
contains a popular exposition of the system, extensive
tables of equivalents for ready reference, a comprehensive
dose table, and over three hundred practical illustrations of
metric prescription writing. The latter are selected from
the pharmacopœias and formularies of the great hospitals
of New York, Philadelphia, Boston, London, etc., or are
contributed from the practice of Medical Officers of the
Marine Hospital Service, who have been using the metric
system exclusively since April 27, 1878. Quite a number
of the prescriptions are transcribed from the Hospital

Formulary, compiled by Charles Rice, Ph. D., of New York. Many of these prescriptions are in fact extensively employed in hospital and out-patient practice.

That the book may prove acceptable and useful is the earnest wish of

THE AUTHOR.

WASHINGTON, D. C.,
February, 1881.

TABLE OF CONTENTS.

PART I.

THE METRIC SYSTEM.

THE METRIC SYSTEM.

To the Honorable John Sherman, our Secretary of the
Treasury, whose practical and clear mind we have to thank
for so many other substantial and lasting benefits, belongs
the honor of being the first representative of our government who has officially authorized and directed the exclu-
sive use of the metric system of weights and measures in
any branch of the public service. Professor C. Lewis Diehl,
in his annual "Report (to the American Pharmaceutical
Association) on the Progress of Pharmacy," for 1879, says
in reference to it: "The action of the Treasury Department in this connection must be regarded as the first practical step to the adoption of the metric system of measure-
ments for all purposes whatever throughout the land."

The Hon. J. K. Upton also—at the time Chief Clerk of
the Treasury Department, and now Assistant Secretary of
the Treasury—made an able and elaborate report upon the
subject of the proposed introduction in this country of the
metric system, in which he took strong ground in favor of
its adoption for use at some fixed early date. The conclud-
ing paragraphs of his report are as follows:

"It should be borne in mind that the only legalized system of weights and measures in this country to-day is the
metric system, and that this system is the only one we
possess in harmony with that of any other country.

16   THE METRIC SYSTEM.

"As to the time necessary for the government and the people to prepare for its obligatory use, there may be some diversity of opinion. Considering the experiences of other nations and the admitted aptness of our people for adopting and utilizing improved methods of business, I am clearly of the opinion that a notice of two years will be sufficient to enable the government to prepare for the adoption of the system in all administrative transactions, and that a notice of ten or fifteen years will be sufficient to enable the country to prepare for its obligatory use in transactions between individuals. Possibly, for a while thereafter, a compromise with vulgar fractions and existing terms may be necessary, but meanwhile the new system will be taught in our schools, explained in the public press, and exemplified by our experience, and in a comparatively brief time the use and terms of the old system will disappear as have those of English money before the advance of our decimal coinage."

It is undeniable, that since the promulgation, with Secretary Sherman's approval, of the order of April 27th, 1878, issued by the late Surgeon-General John M. Woodworth, of the Marine Hospital Service, making the use of the metric system compulsory in that service for all medical and pharmacal purposes, the introduction of this system has made much more rapid strides than ever before among the medical profession, whether this progress be due to that order or not.

The indorsement of the metric system by the American Medical Association, and by a large number of State Medical Associations and County Medical Societies, and especially in the forthcoming new Pharmacopœia of the United States, as directed by the Sixth Decennial Pharmacopœial Convention, have placed the ultimate and complete adoption of that system by the physicians and pharmacists of this country beyond peradventure, though much yet remains to be done to learn the new and forget the old.

How to Learn.

To drop the old system of weights and measures entirely, and start out anew with the metric system, after learning the doses of medicines over again in metric terms, I conceive to be more convenient than safe. During the period of transition from the old to the new system it is necessary to constantly make comparisons between the two, and to convert values as expressed in terms of the one system into their equivalent expressions according to the other. To be on the safe side we will do well to verify the metric doses with which we are making ourselves familiar, by comparison with the doses expressed in the old system about which we have no doubt. Besides, the most thorough way to learn the metric system and its practical application in prescription-writing is to learn the equivalents of the *units* of either system in the units of the other, by heart; to apply convenient rules of conversion, verifying the results by reliable tables; to use a posological table in which the doses are given in terms of both systems, thus enabling the physician to verify the metric dose by direct comparison with that in apothecaries' weights and measures after correcting or modifying the latter dose by the light of his experience and the requirements of the case in hand; and finally to examine already prepared metric prescriptions, such as afford abundant illustrations of familiar practice. Among those given in this volume will be recognized many time-honored formulæ, such as "Stokes's Expectorant," "Scudamore's Gout Mixture," etc. Any physician into whose hands this formulary may come will of course scrutinize each formula which he may wish to use. The special therapeutic application to which each prescription may be put is, for obvious reasons, not given.

Much of the matter which follows will be recognized as substantially the same as prepared by me to accompany the order of April 27th, 1878, above referred to. It is, however, much enlarged and, I think, improved upon. Tables giving exact equivalents, computations of prices, relations of weight to volume, and volume to weight, etc., are added.

Believing that as liquid medicines are necessarily administered in doses by measure, they should also be prepared and prescribed by measure, I have retained volumetric methods in the statement of formulæ. The supposition that the sacrifice of our convenient medicinal fluid measures necessarily follows the adoption of the metric system is erroneous. There can be no doubt that the adoption of the term " Fluigram " for the *cubic-centimeter*, to make plain the analogy between that and the *gram*, as suggested by Mr. Alfred B. Taylor, of Philadelphia, would greatly aid the introduction and usefulness of the metric system, and it is used throughout this formulary with the earnest hope that the innovation will be kindly received. (See New York *Medical Record*, August 14th, 1880.)

Pharmacists, it is taken for granted, will provide themselves with an outfit of metric weights and measures, when the new Pharmacopœia shall have been issued, if they have not already purchased them. There is no longer any excuse for the absence of metric weights and measures in any pharmaceutical establishment. The possession and use of them is moreover the easiest as well as the most thorough method of becoming practically familiar with the system.

THE FRENCH DECIMAL SYSTEM OF WEIGHTS AND MEASURES.

DESCRIPTION.

1. The Metric System is based upon the METER, which is the standard unit of *linear measurement* of that system.

The METER is equal to 39.370432 inches, or about 10 per cent. longer than the yard.

2. The Metric unit of *fluid measure* is the LITER—the cube of $\frac{1}{10}$ meter (one decimeter), or 1000 cubic-centimeters.

The LITER is equal to about 34 fluidounces.

The CUBIC-CENTIMETER is equal to 16.231 minims.

3. The Metric unit of *weight* is the GRAM, which represents the weight of one cubic-centimeter of water at its maximum density.

The GRAM is equal to 15.43234874 troy-grains.

A PRACTICAL VIEW OF IT.

4. In writing and dispensing prescriptions it is always sufficiently accurate and safe to consider 1 Gram exactly equal to 15 troy grains, and 1 Cubic-centimeter as exactly equivalent to 15 minims, and the fractions over 15 may be ignored.

[The actual excess of 1 Gram over 15 troy grains amounts to less than three grains for every hundred grains; and the excess of 1 Cubic-centimeter over 15 minims amounts to $8\frac{1}{3}$ minims for every 100 minims.]

We will accordingly have :

1 Gram equal to $1\frac{5}{1}$ troy-grains

1 Troy grain equal to $\frac{1}{15}$ gram

1 Cubic-centimeter equal to $1\frac{5}{1}$ minims

1 Minim equal to $\frac{1}{15}$ cubic-centimeter.

THE "FLUIGRAM."

5. We have already seen that the Gram is the weight of one Cubic-centimeter of water. Hence, when referring to liquids, the Gram and the Cubic-centimeter may be considered as equal quantities, except the liquids be very heavy (as in the case of chloroform, sulphuric acid, strong solutions, syrups, etc.), or very light (as in the case of ether).

We will hereafter drop the term *Cubic-centimeter*, and adopt in its place the term FLUIGRAM,* to bear in mind the more readily its close relation to the GRAM. The word FLUIGRAM is besides more convenient, euphonious, and American, than the word Cubic-centimeter.

6. The relation between the units of the old system and the units of the Metric System, then, may be expressed as follows :

1 Drachm is equal to 4 Grams; and

1 Fluidrachm is equal to 4 Fluigrams.

RULES FOR CONVERSION.

APOTHECARIES TO METRIC, AND METRIC TO APOTHECARIES.

7. To convert *troy-grains* into Grams, or *minims* into FLUIGRAMS :

A.—*Divide by* 10, *and from the quotient subtract one-third.*

Ex.—To convert 60 grains into its equivalent in Grams, divide the number by 10, which gives 6 as the quotient; from this 6 subtract one-

---

* Proposed by Mr. A. B. Taylor, of Philadelphia.

third (2), which leaves 4 as the answer, 4 Grams being equivalent to 60 grains. Or,

B. *Divide by* 15.

Ex.—3 Fluigrams are equivalent to 45 minims, because 15 is contained in 45 three times.

8. To convert apothecaries' (troy) *drachms* into GRAMS, or *fluidrachms* into FLUIGRAMS:

Multiply by 4.

9. To convert GRAMS into *grains* or FLUIGRAMS into *minims* :

Add 50 *per cent. (one-half), and multiply the sum by* 10.

Ex.—To 12 Grams add 50 per cent. (6), which gives the sum of 18; multiply this by 10, which gives 180 as the number of grains equivalent to 12 Grams.

10. To convert GRAMS into *drachms*, or FLUIGRAMS into *fluidrachms :*

Divide by 4.

11. When whole troy-ounces or fluidounces are concerned, 30 is the multiplicator or divisor to use.

To convert *troy-ounces* (or *fluidounces*) into GRAMS (or FLUIGRAMS):

Multiply by 30; and

To convert GRAMS (or FLUIGRAMS) into *troy-ounces* (or *fluidounces) :*

Divide by 30.

ACCURACY OF THE ABOVE RULES.

12. In applying the rules given in paragraphs 7 to 10 to convert the several quantities by weight in any one prescription or formula constructed exclusively in proportions by weight, *the original proportions between these quantities will still be preserved,* the deviation from exactness being invariably the same. It will also be found that the same rules may be applied in converting the several quantities in a formula

or prescription constructed exclusively by measure, *without disturbing at all the original proportions.*

In any formula, however, where weights and measures are used together, the application of the rules referred to will result in a slight change, so that in the resulting metric formula the measured quantities will be about 5 per cent. larger in proportion to the weighed quantities. This is a fortunate circumstance, because the diluents used in mixtures are the only ones proportionally increased, simply diluting the mixture 5 per cent., which is insignificant and on the safe side.

ABBREVIATIONS.

13. The term Gram is abbreviated "Gm," and the term Cubic-centimeter, "C.C." The word Fluigram is abbreviated "fGm."

PRESCRIPTION WRITING.

13 *a.* In writing prescriptions it is best not to use any other units but the Gm and the fGm, fractions of which should be stated decimally. It is easier in prescriptions to write " 0.10 Gm " than " 10 centigrams," or " 10 cents."

To preclude the possibility of mistaking the sign "Gm" (always written with a capital " G ") for the sign " gr." (which is always written with a small " g "), we should invariably in writing metric prescriptions put the number before the sign, using the common Arabic numerals, to distinguish sufficiently the metric formula from one in apothecaries' weights and measures, where the quantities are written by means of signs, all of which differ widely from " Gm " and " fGm," except the " gr.," which, however, is invariably put in front of the number, which is expressed in Roman numerals.

14. It has been recommended to use a " *decimal line* " instead of decimal points in writing metric prescriptions. This is entirely practicable when the prescription blanks are expressly printed for that purpose, or where the decimal line is drawn from top to bottom. In *writing* a metric prescription, however, in which only one quantity is stated, a short decimal line drawn with the pen would probably be mistaken for the figure " 1."

DIMES, CENTS, AND MILLS.

15. Tenths, hundredths, and thousandths, of both the Gram and Fluigram, may be conveniently called "dimes," "cents," and "mills" when referred to in speaking. The Fluidime being the equivalent of $1\frac{1}{2}$ minim, we will not require any smaller unit to express quantities by measure. In referring to quantities by weight, however, it will be found convenient to use the expression "dime" for 0.10 Gm (equal to $1\frac{1}{2}$ grain), "cent" for 0.01 Gm (equal to about $\frac{1}{6}$ grain), and "mill" for 0.001 Gm (equal to $\frac{1}{64}$ grain). The "cent" and "mill" are especially valuable terms and will be frequently used in this book.

PREFIXES SUPERFLUOUS.

16. The prefixes used in the metric system are simply numerals, as follows:

Myria, which means,		.	.	.	.	.	.	10,000
Kilo, " "		.	.	.	.	.	.	1,000
Hecto, " "		.	.	.	.	.	.	100
Deka, " "		.	.	.	.	.	.	10
	and							
Deci, which means,		.	.	.	.	.	.	0.1
Centi, " "		.	.	.	.	.	.	0.01
Milli " "		.	.	.	.	.	.	0.001

And they are quite unnecessary in the writing of prescriptions (if not in all cases). For multiples of the units Gm and fGm, English numerals are far more convenient, generally intelligible, and at least equally explicit; while for subdivisions of these units, the American terms *dime*, *cent*, and *mill*, are quite handy.

SOMETHING TO LEARN BY HEART.

17. The following approximate equivalents should be *learned by heart:*

1 Troy-grain (or minim) is equal to 0.06 Gram (or Fluigram) or 6 cents.

1 Drachm (or fluidrachm) is equal to 4 Grams (or Fluigrams).

1 Ounce (or fluidounce) is equal to 30 Grams (or Fluigrams).

1 Gram (or Fluigram) is equal to 15 grains (or minims).

4 Grams (or Fluigrams) is equal to 1 drachm (or fluidrachm).   ✦

30 Grams (or Fluigrams) is equal to 1 ounce (or fluidounce).

DROPS AND SPOONFULS.

18. The average " DROP " is equal to 0.05 fGm (or one-half " fluidime "), and 1 fluidime is about 2 drops.

An average TEASPOON holds 5 fGm ; a DESSERTSPOON, 10 fGm ; a TABLESPOON, 20 fGm ; and a WINEGLASS, 75 fGm.

Wherever " 5 fGm " is stated to be the dose of any preparation embraced in this formulary, it is understood that a " teaspoonful " is meant, etc., and that the label should so indicate. This is done so as to show the relation of the dose to the total quantity prescribed. It will, however, also serve to fix in our memory the fact that the average teaspoon holds more than one fluidrachm (4 fGm), and that it is safer and more correct to allow 5 fGm to the teaspoonful in making up the volume of a mixture.

19. TABLES OF APPROXIMATE EQUIVALENTS.

A.

APOTHECARIES' WEIGHTS (AND MEASURES).	METRIC WEIGHTS (AND MEASURES).
Troy-grains (or minims).	*Grams (or Fluigrams).*
$\frac{1}{64}$	0.001 [1 mill, or $\frac{1}{1000}$]
$\frac{1}{32}$	0.002 [2 mills, or $\frac{2}{1000}$]
$\frac{1}{16}$	0.004 [4 mills, or $\frac{4}{1000}$]
$\frac{1}{8}$	0.008 [8 mills, or $\frac{8}{1000}$]
$\frac{1}{4}$	0.016 [16 mills, or 1⅔ cents, or $\frac{16}{1000}$]
$\frac{1}{2}$	0.033 [33 mills, or 3⅓ cents or $\frac{1}{30}$]
1	0.066 [66 mills, or 6⅔ cents or $\frac{1}{15}$]

Apothecaries' Weights (and Measures).	Metric Weights (and Measures.)
Troy-grains (or Minims).	*Grams (or Fluigrams).*
1½	0.100 [10 cents, or $\frac{1}{10}$]
2	0.133 [13⅓ cents, or $\frac{2}{15}$]
3	0.200 [20 cents, or $\frac{1}{5}$]
5	0.333 [33⅓ cents, or ⅓]
6	0.400 [40 cents, or ⅖]
10	0.666 [66⅔ cents, or ⅔]
15	1.000 [1 Gm.]
20	1.333 [1⅓ Gm.]
30	2.000 [2 Gm.]

Drachms (or Fluidrachms).	*Grams (or Fluigrams).*
1	4
2	8
4	16
6	24

Troy-ounces (or Fluidounces).	*Grams (or Fluigrams)*
1	30
2	60
3	90
4	120
5	150
6	180
8	240
10	300
12	360
16	480
20	600

B.

Metric Weights (and Measures). Grams or (Fluigrams).	Apothecaries' Weights (and Measures). Grains (or Minims).
0.001 [1 mill]	1/60
0.002 [2 mills]	1/30
0.003 [3 mills]	1/20
0.004 [4 mills]	1/16
0.005 [5 mills]	1/12
0.006 [6 mills]	1/10
0.008 [8 mills]	1/8
0.010 [10 mills, or 1 cent]	1/6
0.015 [1½ cents, or 15 mills]	1/4
0.020 [2 cents]	1/3
0.03 [3 cents]	1/2
0.04 [4 cents]	2/3
0.05 [5 cents]	3/4
0.065 [6½ cents]	1
0.08 [8 cents]	1 1/4
0.10 [10 cents]	1 1/2
0.15 [15 cents]	2 1/3
0.20 [20 cents]	3
0.30 [30 cents]	4 2/3
0.65 [65 cents]	10
1.00	15
2	30
3	45
	Drachms (or Fluidrachms).
4	1
5	1 1/4
6	1 1/2
7	1 3/4
8	2
9	2 1/4
10	2 1/2
15	4

Metric Weights (and Measures). Grams (or Fluigrams).	Apothecaries' Weights (and Measures). Ounces (or Fluidounces).
30	1
50	1⅔
100	3⅓
250	8⅓
500	17
1000	34

THE METRIC SYSTEM RATIONAL.

20. The essential characteristics of a system of weights and measures which may entitle it to cosmopolitan adoption assuperior to the old arbitrary and incongruous systems, or rather no-systems, are: 1st. That it rest upon a basis ot some geographical magnitude ; 2d. That it be a decimal system ; 3d. That the unit for linear measurement be the primary unit of the whole system, to which the units for measurement of surface, volume, and weight, derived from it, shall bear the simplest relation possible. These conditions are fulfilled by the metric system based upon the meter.

37. The following tables of equivalents are exact and reliable. They are all based entirely upon the determination of the length of the Meter in English inches (by Captain Clarke), and the determination of the value of the Gram in troy-grains (by Professor Miller).

THE EXACT EQUIVALENTS OF SOME UNITS OF WEIGHT AND MEASURE.

1 Yard . . . . . =	0.914392 Meter.
1 " . . . . . =	91.439178 Centimeters.
1 Foot . . . . . =	0.304797 Meter.
1 " . . . . . =	30.479726 Centimeters.
1 Inch . . . . . =	0.025400 Meter.
1 " . . . . . =	2.539977 Centimeters.
1 " . . . . . =	25.39977 Millimeters.

1 Meter . . . . . =	1.093623 Yard.
1 " . . . . . =	3.280869 Feet.
1 " . . . . . =	39.370432 Inches.
1 Centimeter . . . . =	0.393704 "
1 Millimeter . . . . =	0.039370 "

1 Wine Gallon . . . =	3.785310 Liters.
1 " " . . . . =	3785.309889 Cubic-centimeters.
1 Wine Quart . . . . =	0.94632715 Liter.
1 " " . . . . =	946.327472 Cubic-centimeters.
1 Wine Pint . . . . =	0.473164 Liter.
1 " " . . . . =	473.163736 Cubic-centimeters.
1 U. S. Fluidounce . . . =	29.572734 " "
1 U. S. Fluidrachm . . . =	3.696592 " "
1 U. S. Minim . . . . =	0.061609 " "
1 " " . . . . =	0.616086 Fluidimes.
1 Liter . . . . . =	0.264179 Wine Gallon.

I Liter . . . . . =	1.056717 Wine Quart.
I " . . . . . =	2.113433 Wine Pints.
I " . . . . . =	33.814933 U. S. Fluidounces.
I " . . . . . =	270.519463 U. S. Fluidrachms.
I Cubic-centimeter (Fluigram) =	16.231169 U. S. Minims.
I Fluidime (0.1 cc.) . . =	1.623117 " "

I Cubic Inch . . . . =	0.016387 Liters.
I " " . . . . =	16.386623 Cubic-centimeters.

I Liter . . . . . =	61.025387 Cubic Inches.
I Cubic-centimeter (Fluigram) . =	0.061025 " "

I Imperial Gallon . . . =	4.540922 *Liters.
I " " . . . =	4540.921544 Cubic-centimeters.
I Imperial Quart . . . =	1.135230 Liter.
I " " . . . =	1135.230386 Cubic-centimeters.
I Imperial Pint . . . =	0.567615 Liter.
I " " . . . =	567.615193 Cubic-centimeters.
I Imperial Fluidounce . . =	28.380760 " "
I Imperial Fluidrachm . . =	3.547595 " "
I Imperial Minim . . . =	0.059127 " "

I Liter . . . . . =	0.220220 *Imperial Gallon.
I " . . . . . =	0.880878 Imperial Quart.
I " . . . . . =	1.761757 Imperial Pint.

* One Imperial Gallon being the volume of 70,000 grains of pure water at + 62° F. (+ 16.66° C.), and the specific gravity of water at

1 Liter, ... =	35.235139	Imp'l Fluidounces.
1 " . . . . . =	281.881109	Imp'l Fluidrachms.
1 Cubic-centimeter (Fluigram) =	16.912866	Imperial Minims.
1 Fluidime (0.1 cc.) . . . =	1.691287	" "

+62° F. being assumed to be 0.9989, the number of Cubic-centimeters in an Imperial Gallon has been obtained as follows :

$$\frac{4535.9265254655 \ (= \text{the number of grams equal to } 70,000 \text{ grains}).}{.9989}$$

One Liter of water at +4° C. weighs 15,432.34874 grains and at +62° F. 15,415.373156386 grains (assuming the Specific Gravity of water at 62° F. to be 0.9989), while 15,415.373156386 grains of water +62° F. measures 0.220219616 Imperial Gallon : $= \frac{15,415.373156386}{70,000}$

As 1000 Cubic-centimeters is equal to 61.02538677102464126 1568 cubic inches (assuming as correct the determination by Captain Clarke of the value of the Meter in English inches = 39.370432), and as 1000 Cubic-centimeters of pure water at +4° C. (barometer 30 inches) weighs 15,432.34874 troy-grains (assuming as correct the determination by Professor Miller of the value of the Gram in troy-grains = 15.43234874), the weight of 1 cubic inch of pure water at + 4° C. should be 252.88408 troy grains, and at +16.66° C. (= 62° F.) $\frac{252.88408 \times 0.9989}{1.0000}$ = 252.6059 grains.

If the number of cubic-centimeters in an Imperial Gallon, obtained as stated above (by dividing the number of grams equal to 70,000 grains by .9989), be correct, then, as 1 cubic-centimeter is equal to 0.061025386, etc., cubic inches, each Imperial Gallon contains 4,540,-921544 × 0.061025386 = 277.1115 cubic inches.

Both the above results are verified by multiplying the number of cubic inches contained in one Imperial Gallon (277.1115) by the weight of 1 cubic inch of pure water at 62° F. (252.6059), which will give the number of grains of pure water at + 62° the volume of which is the Imperial Gallon (70,000). The result of this multiplication is in fact 69,999.9998578, which proves the correctness of the calculations.

1 Wine Gallon	=	0.833599 Imperial Gallon.
1 " "	=	3.334398 Imperial Quarts.
1 " "	=	6.668796 Imperial Pints.
1 " "	=	133.375915 Imperial Fluidounces.
1 " "	=	1,067.007322 Imperial Fluidrachms.
1 " "	=	64,020.439426 Imperial Minims.
1 Wine Quart	=	0.833599 Imperial Quart.
1 " "	=	1.667199 Imperial Pints.
1 " "	=	33.343978 Imperial Fluidounces.
1 " "	=	266.751831 Imperial Fluidrachms.
1 " "	=	16,005.109856 Imperial Minims.
1 Wine Pint	=	0.833599 Imperial Pint.
1 " "	=	16.671989 Imperial Fluidounces.
1 " "	=	133.375915 Imperial Fluidrachms.
1 " "	=	8,002.554928 Imperial Minims.
1 U. S. Fluidounce	=	1.041999 Imperial Fluidounces.
1 " "	=	8.335995 Fluidrachms.
1 " "	=	500.159683 Imperial Minims.
1 U. S. Fluidrachm	=	1.041999 Imperial Fluidrachms.
1 " "	=	62.519960 Imperial Minims.
1 U. S. Minim	=	1.041999 " "

_____

1 Imperial Gallon	=	1.199617 Wine Gallons.
" "	=	4.798468 Wine Quarts.
" "	=	9.596935 Wine Pints.
" "	=	153.550961 U. S. Fluidounces.
" "	=	1,228.407688 U. S. Fluidrachms.
" "	=	73,704.46129 U. S. Minims.
1 Imperial Quart	=	1.199617 Wine Quarts.
1 " "	=	2.399234 Wine Pints.
1 " "	=	38.387740 U. S. Fluidounces.
1 " "	=	307.101922 U. S. Fluidrachm.
1 " "	=	18,426.115323 U. S. Minims.
1 Imperial Pint	=	1.199617 Wine Pints.
1 " "	=	19.193870 U. S. Fluidounces.
1 " "	=	153.550961 U. S. Fluidrachms.
1 " "	=	9,213.057661 U. S. Minims.

1 Imperial Fluidounce	. =	.959694 U. S. Fluidounces.
1 " "	. . =	7.677548 U. S. Fluidrachms.
1 " "	. . =	460.652883 U. S. Minims.
1 Imperial Fluidrachm	. =	0.959694 U. S. Fluidrachms.
1 " "	. . =	57.581610 U. S. Minims.
1 Imperial Minim	. . =	0.959694 " "

---

1 Troy Pound	. . . =	373.242954 Grams.
1 Troy Ounce	. . . =	31.103496 "
1 Troy Drachm	. . =	3.887937 "
1 Scruple	. . . . =	1.295979 "
1 Troy Grain	. . . =	0.064799 "
1 " "	. . . =	6.479895 Centigrams.
1 " "	. . =	64.798950 Milligrams.

---

1 Kilogram	. . . =	2.679227 Troy Pounds.
1 "	. . . . =	32.150727 Troy Ounces.
1 "	. . . . =	257.205812 Troy Drachms.
1 "	. . . . =	771.617437 Scruples.
1 "	. . . . =	15,432.348740 Troy Grains.
1 Gram	. . . =	15.432349 "
1 Decigram (Dime)	. . =	1.543235 "
1 Centigram (Cent)	. . =	.154323 "
1 Milligram (Mill)	. . =	.015432 "

---

1 Avoirdupois Pound	. =	453.592653 Grams.
1 Avoirdupois Ounce	. =	28.349541 "
1 Avoirdupois Drachm	. =	1.771846 "
1 Avoirdupois Grain	. . =	.064799 "
1 Kilogram	. . =	2.204621 Avoirdupois Pounds.
1 "	. . . =	35.273940 Avoirdupois Ounces.
1 "	. . . =	564.383040 Avoirdupois Drachms.
1 "	. . . =	15,432.348740 Avoirdupois Grains.

1 Troy Pound	. . . =	0.822857	Avoirdupois Pound.
1 "	. . . =	13.165714	Avoirdupois Ounces.
1 "	. . . =	210.651429	Avoirdupois Drachms.
1 Troy Ounce	. . . =	1.097143	Avoirdupois Ounces.
1 "	. . . =	17.554286	Avoirdupois Drachms.
1 Troy Drachm	. . . =	2.194286	" "
1 "	. . . =	60.000000	Avoirdupois Grains.

---

1 Avoirdupois Pound	. =	1.215278	Troy Pounds.
1 " "	. =	14.583333	Troy Ounces.
1 " "	. =	11.666667	Troy Drachms.
1 " "	. =	35.000000	Troy Scruples.
1 Avoirdupois Ounce	. =	0.911458	Troy Ounces.
1 " "	. =	7.291667	Troy Drachms.
1 " "	. =	21.875000	Scruples.
1 " "	. =	437.500000	Grains.
1 Avoirdupois Drachm	. =	0.455729	Troy Drachms.
1 " "	. =	1.367187	Scruples.
1 " "	. =	27.343750	Grains.

MORE TABLES OF EQUIVALENTS.

A.—Relation of Metric to English Measures of Length.

(1 meter = 39.370432 inches.—*Clarke.*)

Meters.	Equivalents in			Meters.	Equivalents in		
	Inches.	Feet.	Yards.		Inches.	Feet.	Yards.
0.001	0.039			11	433.075	36.099	12.030
0.010	0.394			12	472.445	39.370	13.123
0.100	3.937	0.328	0.109	13	511.816	42.651	14.217
1.000	39.370	3.281	1.094	14	551.186	45.932	15.311
2	78.741	6.562	2.187	15	590.556	49.213	16.404
3	118.111	9.843	3.281	16	629.927	52.494	17.498
4	157.482	13.123	4.374	17	669.297	55.775	18.592
5	196.852	16.404	5.468	18	708.668	59.056	19.685
6	236.223	19.685	6.562	19	748.038	62.337	20.779
7	275.593	22.966	7.655	20	787.409	65.617	21.872
8	314.963	26.247	8.749	100	3937.043	328.087	109.362
9	354.334	29.528	9.843	1000	3,9370.432	3280.869	1093.623
10	393.704	32.809	10.936				

[The above table can also be used to ascertain the value of any article per meter, the price per inch, foot, or yard being known. To find it, let the figures in the first column, under the head of "Meters," represent the known price, and the figures in the other columns the prices sought. *Ex.*—If the price is 2 cents per inch, find 2 in the first column, and opposite in the inch column will then be found the price per meter, viz., 78¾ cents. At 2 cents per foot, the meter would cost 6.56 cents; and at 10 cents per yard the meter would cost 10.94 per yard, etc. See also footnote to Table E.]

B.—RELATION OF ENGLISH TO METRIC MEASURES OF LENGTH.

(1 yard = 0.91439178 meters.)

1 inch	. . = 0.025 meters.	17 feet .	. =	5.182 meters.
2 "	. . = 0.051 "	18 " .	. =	5.486 "
3 "	. . = 0.076 "	19 ' .	. =	5.791 "
4 "	. . = 0.102 "	20 " .	. =	6.096 "
5 "	. . = 0.127 "	21 " .	. =	6.401 "
6 "	. . = 0.152 "	22 " .	. =	6.706 "
7 "	. . = 0.178 "	23 " .	. =	7.010 "
8 "	. . = 0.203 "	24 " .	. =	7.315 "
9 "	. . = 0.229 "	25 " .	. =	7.620 "
10 "	. . = 0.254 "	26 " .	. =	7.925 "
11 "	. . = 0.279 "	27 " .	. =	8.230 "
1 foot	. . = 0.305 "	28 " .	. =	8.534 "
2 feet	. . = 0.610 "	29 " .	. =	8.839 "
3 "	. . = 0.914 "	30 " .	. =	9.144 "
4 "	. . = 1.219 "	31 " .	. =	9.449 "
5 "	. . = 1.524 "	32 " .	. =	9.754 "
6 "	. . = 1.829 "	33 " .	. =	10.058 "
7 "	. . = 2.134 "	34 " .	. =	10.363 "
8 "	. . = 2.438 "	35 " .	. =	10.668 "
9 "	. . = 2.743 "	36 " .	. =	10.973 "
10 "	. . = 3.048 "	37 " .	. =	11.277 "
11 "	. . = 3.353 "	38 " .	. =	11.582 "
12 "	. . = 3.658 "	39 " .	. =	11.887 "
13 "	. . = 3.962 "	40 " .	. =	12.192 "
14 "	. . = 4.267 "	50 " .	. =	15.24 "
15 "	. . = 4.572 "	100 " .	. =	30.480 "
16 "	. . = 4.877 "			

[The above table will also show the value per inch or foot of any article the price of which is given per meter. See footnote to previous table and to Table E.]

(1 square meter is equal to 1550.030915870 square inches.)

C.—Relation of Metric to U. S. Fluid Measures.

(1 cubic-centimeter = 16.2311678 + minims.)

0.05 cc.	.	. = 0.81 + min.	2 cc. .	. =	32.46 + min.
0.06 "	.	. = 0.97 + "	3 " .	. =	48.69 + "
0.07 "	.	. = 1.14 — "	4 " .	. =	1.08 + fl. drs.
0.08 "	.	. = 1.30 — "	5 " .	. =	1.35 + "
0.09 "	.	. = 1.46 + "	6 " .	. =	1.62 + "
0.10 "	.	. = 1.62 + "	7 " .	. =	1.89 + "
0.11 "	.	. = 1.79 — "	8 " .	. =	2.16 + "
0.12 "	.	. = 1.95 — "	9 " .	. =	2.43 + "
0.13 "	.	. = 2.11 + "	10 " .	. =	2.71 — "
0.14 "	.	. = 2.27 + "	20 " .	. =	5.41 + "
0.15 "	.	. = 2.43 + "	30 " .	. =	1.01 + fl. ozs.
0.16 "	.	. = 2.60 — "	40 " .	. =	1.35 + "
0.17 "	.	. = 2.76 — "	50 " .	. =	1.69 + "
0.18 "	.	. = 2.92 + "	60 " .	. =	2.03 — "
0.19 "	.	. = 3.08 + "	70 " .	. =	2.37 — "
0.20 "	.	. = 3.25 — "	80 " .	. =	2.71 — "
0.25 "	.	. = 4.06 — "	90 " .	. =	3.04 + "
0.30 "	.	. = 4.87 + "	100 " .	. =	3.38 + "
0.35 "	.	. = 5.68 — "	150 " .	. =	5.07 + "
0.40 "	.	. = 6.49 + "	200 " .	. =	6.76 + "
0.45 "	.	. = 7.30 + "	250 " .	. =	8.45 + "
0.50 "	.	. = 8.12 — "	300 " .	. =	10.14 + "
0.55 "	.	. = 8.93 — "	350 " .	. =	11.84 — "
0.60 "	.	. = 9.74 — "	400 " .	. =	13.53 — "
0.65 "	.	. = 10.55 + "	450 " .	. =	15.22 — "
0.70 "	.	. = 11.36 + "	500 " .	. =	1.06 — pints.
0.75 "	.	. = 12.17 + "	600 " .	. =	1.27 — "
0.80 "	.	. = 12.98 + "	700 " .	. =	1.45 — "
0.85 "	.	. = 13.80 — "	800 " .	. =	1.69 + "
0.90 "	.	. = 14.61 — "	900 " .	. =	1.90 + "
0.95 "	.	. = 15.42 — "	1000 " .	. =	2.11 — "
1 "	.	. = 16.23 + "			

[This table also incidentally shows the price per cubic-centimeter of any article the price of which is known per minim, drachm, or fluid-ounce, or pint. See footnotes to Tables A, E, and K.]

D.—Relation of U. S. to Metric Fluid Measures.

(1 minim = 0.061613 + cubic-centimeter.)

1 minim	.	. =	0.06 + cc.	60 min. .	. =	3.70 — cc.
2 "	.	. =	0.12 + "	70 " .	. =	4.31 + "
3 "	.	. =	0.18 + "	80 " .	. =	4.93 — "
4 "	.	. =	0.25 — "	90 " .	. =	5.54 + "
5 "	.	. =	0.31 — "	100 " .	. =	6.16 + "
6 "	.	. =	0.37 — "	110 " .	. =	6.78 — "
7 "	.	. =	0.43 + "	120 " .	. =	7.39 + "
8 "	.	. =	0.49 + "	3 fl. drs. .	. =	11.09 — "
9 "	.	. =	0.55 + "	4 " .	. =	14.79 — "
10 "	.	. =	0.62 — "	5 " .	. =	18.48 + "
11 "	.	. =	0.68 — "	6 " .	. =	22.18 — "
12 "	.	. =	0.74 — "	7 " .	. =	25.88 — "
13 "	.	. =	0.80 + "	8 " .	. =	29.57 + "
14 "	.	. =	0.86 + "	9 " .	. =	33.27 — "
15 "	.	. =	0.92 + "	10 " .	. =	36.97 — "
16 "	.	. =	0.99 — "	11 " .	. =	40.66 + "
17 "	.	. =	1.05 — "	12 " .	. =	44.36 — "
18 "	.	. =	1.11 — "	13 " .	. =	48.06 — "
19 "	.	. =	1.17 + "	14 " .	. =	51.75 + "
20 "	.	. =	1.23 + "	15 " .	. =	55.45 — "
21 "	.	. =	1.29 + "	16 " .	. =	59.10 — "
22 "	.	. =	1.36 — "	3 fl. ozs. .	. =	88.67 — "
23 "	.	. =	1.42 — "	4 " .	. =	118.24 + "
24 "	.	. =	1.48 — "	5 " .	. =	147.81 + "
25 "	.	. =	1.54 + "	6 " .	. =	177.39 — "
26 "	.	. =	1.60 + "	7 " .	. =	206.96 — "
27 "	.	. =	1.66 + "	8 " .	. =	236.53 + "
28 "	.	. =	1.73 — "	9 " .	. =	266.10 + "
29 "	.	. =	1.79 — "	10 " .	. =	295.68 — "
30 "	.	. =	1.85 — "	11 " .	. =	325.25 + "
35 "	.	. =	2.16 — "	12 " .	. =	354.82 + "
40 "	.	. =	2.46 + "	13 " .	. =	384.40 — "
45 "	.	. =	2.77 + "	14 " .	. =	413.97 — "
50 "	.	. =	3.08 + "	15 " .	. =	443.54 + "
55 "	,	. =	3.39 — "	16 " .	. =	473.11 + "

D.—*Relation of U. S. to Metric Fluid Measures*—Continued.

17 fl. ozs.	. =	502.69 — cc.	26 fl. ozs.	. =	768.94 + cc.
18 "	. =	532.26 — "	27 "	. =	798.51 + "
19 "	. =	561.93 + "	28 "	. =	828.09 — "
20 "	. =	591.50 + "	29 "	. =	857.66 — "
21 "	. =	621.08 — "	30 "	. =	887.23 — "
22 "	. =	650.65 + "	31 "	. =	916.80 + "
23 "	. =	680.22 + "	32 "	. =	946.38 — "
24 "	. =	709.80 + "	64 "	. =	1892.75 + "
25 "	. =	739.37 — "	128 "	. =	3785.51 — "

[The above table may be used also to ascertain the price per minim, drachm or ounce of any article the price of which is known per cubic centimeter. See footnote to Tables A, E, and F.]

E.—Relation of Metric to Apothecaries' Weights.

(1 gram = 15.43234874 troy grains.)

0.0010 gram	. = 0.015 grain.	0.0125 gram	. = 0.193 grain.
0.0013 "	. = 0.019 "	0.0150 "	. = 0.231 "
0.0015 "	. = 0.023 "	0.0200 "	. = 0.309 "
0.0020 "	. = 0.031 "	0.0250 "	. = 0.386 "
0.0025 "	. = 0.039 "	0.0300 "	. = 0.463 "
0.0030 "	. = 0.046 "	0.0350 "	. = 0.540 "
0.0035 "	. = 0.054 "	0.0400 "	. = 0.617 "
0.0040 "	. = 0.062 "	0.0450 "	. = 0.694 "
0.0045 "	. = 0.069 "	0.050 "	. = 0.772 "
0.0050 "	. = 0.077 "	0.055 "	. = 0.849 "
0.0055 "	. = 0.085 "	0.060 "	. = 0.926 "
0.0060 "	. = 0.093 "	0.065 "	. = 1.003 "
0.0065 "	. = 0.100 "	0.070 "	. = 1.080 "
0.0070 "	. 0.108 "	0.075 "	. = 1.157 "
0.0075 "	. = 0.116 "	0.080 "	. = 1.235 "
0.0080 "	. = 0.123 "	0.085 "	. = 1.312 "
0.0085 "	. = 0.131 "	0.090 "	. = 1.389 "
0.0090 "	. = 0.139 "	0.095 "	. = 1.466 "
0.0095 "	. = 0.147 "	0.100 "	. = 1.543 "
0.0100 "	. = 0.154 "	0.110 "	. = 1.698 "

E.—*Relation of Metric to Apothecaries' Weights*—Continued.

0.120 gram	=	1.852 grs.	5 grams	. =	77.162 grs.	
0.130 "	=	2.006 "	6 "	. =	92.594 "	
0.140 "	=	2.161 "	7 "	. =	108.026 "	
0.150 "	=	2.315 "	8 "	. =	123.459 "	
0.160 "	=	2.469 "	9 "	. =	138.891 "	
0.170 "	=	2.623 "	10 "	. =	154.323 "	
0.180 "	=	2.778 "	11 "	. =	169.756 "	
0.190 "	=	2.932 "	12 "	. =	185.188 "	
0.200 "	=	3.086 "	13 "	. =	200.621 "	
0.210 "	=	3.241 "	14 "	. =	216.053 "	
0.220 "	=	3.395 "	15 "	. =	231.485 "	
0.230 "	=	3.549 "	16 "	. =	246.918 "	
0.240 "	=	3.704 "	17 "	. =	262.350 "	
0.250 "	=	3.858 "	18 "	. =	277.782 "	
0.260 "	=	4.012 "	19 "	. =	293.215 "	
0.270 "	=	4.167 "	20 "	. =	308.647 "	
0.280 "	=	4.321 "	21 "	. =	324.079 "	
0.290 "	=	4.475 "	22 "	. =	339.512 "	
0.300 "	=	4.630 "	23 "	. =	354.944 "	
0.310 "	=	4.784 "	24 "	. =	370.376 "	
0.320 "	=	4.938 "	25 "	. =	385.809 "	
0.330 "	=	5.093 "	26 "	. =	401.241 "	
0.340 "	=	5.247 "	27 "	. =	416.673 "	
0.350 "	=	5.401 "	28 "	. =	432.106 "	
0.360 "	=	5.556 "	29 "	. =	447.538 "	
0.370 "	=	5.710 "	30 "	. =	462.970 "	
0.380 "	=	5.864 "	31 "	. =	478.403 "	
0.390 "	=	6.019 "	32 "	. =	493.835 "	
0.400 "	=	6.173 "	33 "	. =	509.268 "	
0.500 "	=	7.716 "	34 "	. =	524.700 "	
0.600 "	=	9.259 "	35 "	. =	540.132 "	
0.700 "	=	10.803 "	36 "	. =	555.565 "	
0.800 "	=	12.346 "	37 "	. =	570.997 "	
0.900 "	=	13.889 "	38 "	. =	586.429 "	
1 "	=	15.432 "	39 "	. =	601.862 "	
2 "	=	30.865 "	40 "	. =	617.294 "	
3 "	=	46.297 "	50 "	. =	771.617 "	
4 "	=	61.729 "	60 "	. =	925.941 "	
			70 "	. =	1,080.264 "	

E.—*Relation of Metric to Apothecaries' Weights*—Continued.

80 gms.	=	1,234.588 grs.	800 gms.	=	12,345.879 grs.
90 "	=	1,388.911 "	900 "	=	13,889.114 "
100 "	=	1,543.235 "	1,000 "	=	15,432.35 "
125 "	=	1,929.044 "	1,500 "	=	23,148.52 "
150 "	=	2,314.852 "	2,000 "	=	30,864.70 "
200 "	=	3,086.470 "	2,500 "	=	38,580.87 "
250 "	=	3,858.087 "	3,000 "	=	46,297.05 "
300 "	=	4,629.705 "	3,500 "	=	54,013.22 "
333 "	=	5,144.118 "	4,000 "	=	61,729.40 "
350 " =	5,401.322 "	4,500 "	=	69,445 57 "
400 "	=	6,172.940 "	5,000 "	=	77,161.74 "
450 "	=	6,944.557 "	10,000 "	=	154,432.35 "
500 "	=	7,716.174 "	20,000 "	=	308,646.97 "
600 "	=	9,259.409 "	25,000 "	=	385,808.72 "
700 "	=	10,802.644 "	50,000 "	=	771,617.44 "
750 "	=	11,574.262 "	100,000 "	=	1543,234.87 "

[This table also shows the price per Gram of any article the price of which is known by the grain; assuming the price per grain to be represented by the figures in the first column, the price per gram is represented by the figures opposite in the next column.]

F.—RELATION OF APOTHECARIES' TO METRIC WEIGHTS.

(1 troy-grain = 0.06479895 + gram.)

$\frac{1}{64}$ troy-grain	= 0.00101 + gm.	$\frac{1}{10}$ troy-grain	= 0.00648 — gm.
$\frac{1}{60}$ "	= 0.00108 — "	$\frac{1}{8}$ "	= 0.00810 — "
$\frac{1}{50}$ "	= 0.00130 — "	$\frac{1}{6}$ "	= 0.01080 — "
$\frac{1}{48}$ "	= 0.00135 — "	$\frac{1}{5}$ "	= 0.01296 — "
$\frac{1}{40}$ "	= 0.00162 — "	$\frac{1}{4}$ "	= 0.01620 — "
$\frac{1}{36}$ "	= 0.00180 — "	$\frac{1}{3}$ "	= 0.02160 — "
$\frac{1}{32}$ "	= 0.00202 + "	$\frac{1}{2}$ "	= 0.03240 — "
$\frac{1}{30}$ "	= 0.00216 — "	$\frac{3}{4}$ "	= 0.04860 — "
$\frac{1}{25}$ "	= 0.00259 + "	1 "	= 0.06480 — "
$\frac{1}{24}$ "	= 0.00270 — "	1½ "	= 0.09720 — "
$\frac{1}{20}$ "	= 0.00324 — "	2 "	= 0.12960 — "
$\frac{1}{18}$ "	= 0.00360 — "	2½ "	= 0.16200 — "
$\frac{1}{16}$ "	= 0.00405 — "	3 "	= 0.19440 — "
$\frac{1}{15}$ "	= 0.00432 — "	4 "	= 0.25920 — "
$\frac{1}{12}$ "	= 0.00540 — "	5 "	= 0.32399 + "

F.—*Relation of Apothecaries' to Metric Weights*—Continued.

6 troy-grains		0.38879	— gm.	1 ounce	=	31.103 + gm.
7	"	= 0.45359	— "	1½ "	=	46.655 + "
8	"	= 0.51839	— "	2 "	=	62.207 — "
9	"	= 0.58319	+ "	3 "	=	93.310 + "
10	"	= 0.64799	— "	4 "	=	124.414 — "
11	"	= 0.71297	— "	5 "	=	155.517 + "
12	"	= 0.77759	— "	6 "	=	186.621 — "
13	"	= 0.84239	— "	7 "	=	217.724 + "
14	"	= 0.90718	+ "	8 "	=	248.823 — "
15	"	= 0.97198	+ "	9 "	=	279.931 + "
16	"	= 1.037	— "	10 "	=	311.035 — "
17	"	= 1.102	— "	11 "	=	342.138 + "
18	"	= 1.166	+ "	12 "	=	373.250 — "
19	"	= 1.231	+ "	13 "	=	404.345 + "
20	"	= 1.296	— "	14 "	=	435.449 — "
21	"	= 1.361	— "	15 "	=	466.552 + "
22	"	= 1.426	— "	16 "	=	497.656 — "
23	"	= 1.458	— "	17 "	=	528.759 + "
24	"	= 1.555	+ "	18 "	=	559.863 + "
25	"	= 1.620	+ "	19 "	=	590.966 + "
26	"	= 1.685	— "	20 "	=	622.070 — "
27	"	= 1.749	+ "	21 "	=	653.173 + "
28	"	= 1.814	+ "	22 "	=	684.277 — "
29	"	= 1.869	— "	23 "	=	715.380 + "
30	"	= 1.944	— "	24 "	=	746.499 + "
40	"	= 2.592	— "	25 "	=	777.587 + "
50	"	= 3.240	— "	26 "	=	808.691 — "
1 drachm		= 3.888	— "	27 "	=	839.794 + "
2 drachms		= 7.776	— "	28 "	=	870.898 — "
3	"	=11.664	— "	29 "	=	902.001 + "
4	"	=15.552	— "	30 "	=	933.105 — "
5	"	=19.440	— "	40 "	=	1244.14 + "
6	"	=23.328	— "	50 "	=	1555.17 + "
7	"	=27.216	— "	100 "	=	3110.35 — "

[The above table also shows the price per grain, troy-drachm, or troy-ounce, of any article the price of which is known per Gram. Assuming that the figures in the first column represent the price per Gram, the prices per grain, troy-drachm, or troy-ounce, are found opposite. *Ex.*—If the Gram is worth 1½ cents, the grain is worth 0.09 cents. If the Gram is worth 1 cent, the drachm is worth 3.88 cents. If the Gram is worth 5 cents, the troy-ounce is worth $1.55½.]

Relation of Volume to Weight and of Weight to Volume.

At +4° C. (39.2° F.); barometer 30 inches.

PURE WATER.

1 Liter weighs, ... 1 Kilogram.
1 Cubic-centimeter (Fluigram) weighs, . . . 1 Gram.

---

1 Kilogram measures, . . 1 Liter.
1 Gram measures, . . . 1 Cubic-centimeter (Fluigram).

---

1 Liter weighs, ... 15,432.34874 Grains.
1 Cubic-centimeter weighs, . . . 15.43234874 "

---

1 Liter weighs, . . . 2.679227 Troy Pounds.
1 " " . . . 32.150727 Troy Ounces.
1 " " . . . 257.205812 Troy Drachms.
1 " " . . . 2.204621 Avoirdupois Pounds.
1 " " . . . 35.273940 Avoirdupois Ounces.
1 " " . . . 564.383040 Avoirdupois Drachms.

---

1 Wine Gallon weighs, ... 58,416.2222 Grains.
1 Wine Quart weighs, ... 14,604.0554 "
1 Wine Pint weighs, ... 7,302.0277 "
1 U. S. Fluidounce weighs, . . . 456.3767 "
1 U. S. Fluidrachm weighs, . . . 57.0471 "
1 U. S. Minim weighs, ...950785 "

---

1 Wine Gallon weighs, ... 3,785.309888 Grams.
1 Wine Quart weighs, ... 946.327472 "
1 Wine Pint weighs, ... 473.163736 "
1 U. S. Fluidounce weighs, . . . 29.572733 "
1 U. S. Fluidrachm weighs, . . . 3.696592 "
1 U. S. Minim weighs, ... 0.061609 "

1 Imperial Gallon weighs, . . . 70,077.08479 Grains.
1 Imperial Quart weighs, . . . 17,519.27120 "
1 Imperial Pint weighs, . . : 8,759.63560 "
1 Imperial Fluidounce weighs, . . 437.98178 "
1 Imperial Fluidrachm weighs, . . 54.74772 "
1 Imperial Minim weighs, . . . 0.91246 "

---

1 Imperial Gallon weighs, ... 4,540.92154 Grams.
1 Imperial Quart weighs, ... 1,135.23038 "
1 Imperial Pint weighs, ... 567.61519 "
1 Imperial Fluidounce weighs, . . . 28.38076 "
1 Imperial Fluidrachm weighs, . . . 3.54759 "
1 Imperial Minim weighs, . . . 0.05913 "

---

1 Cubic Inch weighs, ... 252.88408 Grains.

---

1 Cubic Inch weighs, ... 16.38662 Grams.

---

1 Troy Pound measures, . . . 6,058.168997 U. S. Minims.
1 Troy Ounce measures, . . . 504.847416 " "
1 Troy Drachm measures, . . 63.105927 " "
1 Scruple measures, . . . 21.035309 " "
1 Grain measures, ... 1.051766 " "

---

1 Troy Pound measures, . . 373.242954 Cubic-centimeters.
1 Troy Ounce measures, . . 31.103496 " "
1 Troy Drachm measures, . . 3.887937 " "
1 Troy Grain measures, . . 0.064799 " "

---

1 Troy Pound measures, . . 6,313.808090 Imperial Minims.
1 Troy Ounce measures, . . 526.150674 " "
1 Troy Drachm measures, . . 65.767584 " "
1 Troy Grain measures, . . 1.096126 " "

1 Avoirdupois Pound measures, . 7,671.551969 Imperial Minims.
1 Avoirdupois Ounce measures, .   474.719980   "   "

1 Avoirdupois Pound measures, . 453.592653 Cubic-centimeters.
1 Avoirdupois Ounce measures, .   28.349541   "   "

1 Avoirdupois Pound measures, . 7,362.337998 U. S. Minims.
1 Avoirdupois Ounce measures, .   460.146125   "   "

At +16.66° C. (+62° F.); barometer 30 inches.

PURE WATER.

1 Liter weighs, ... 0.9989 Kilogram.
1 Cubic-centimeter weighs,   .   . 0.9989 Cubic-centimeters.

1 Kilogram measures, .   .   . 1.0011012 Liters.
1 Gram measures,   .   .   . 1.0011012 Cubic-centimeters.

1 Liter weighs, ... 15,415.373156 Grains.

TABLE OF THERMOMETRIC EQUIVALENTS.

To reduce Centigrade degrees to those of Fahrenheit :
RULE.—Multiply by 9, divide by 5, and add 32.

To reduce Fahrenheit's degrees to those of the Centigrade scale :
RULE.—Subtract 32, multiply by 5, and divide by 9.

Table of Equivalents.

Fahrenheit.	Centigrade or Celsius.	Fahrenheit.	Centigrade or Celsius.	Fahrenheit.	Centigrade or Celsius.
2570	1410	206.6	97	129.2	54
2030	1110	204.8	96	127.4	53
1490	810	203	95	125.6	52
1004	540	201.2	94	123.8	51
932	500	199.4	93	122	50
752	400	200	93.33	120.2	49
572	300	197.6	92	118.4	48
500	260	195.8	91	116.6	47
482	250	194	90	116	46.66
464	240	192.2	89	115	46.11
446	230	190.4	88	114.8	46
428	220	188.6	87	114	45.55
410	210	186.8	86	113	45
392	200	185	85	112	44.44
374	190	183.2	84	111.2	44
365	185	181.4	83	111	43 88
356	180	179.6	82	110	43.33
347	175	177.8	81	109.4	43
338	170	176	80	109	42.77
329	165	174.2	79	108.5	42.50
320	160	172.4	78	108	42.22
311	155	170.6	77	107.6	42
302	150	168.8	76	107	41.66
300	148.88	167	75	106	41.11
293	145	165.2	74	105.8	41
284	140	163.4	73	105	40.55
275	135	161.6	72	104	40
266	130	159.8	71	103	39.46
257	125	158	70	102.2	39
248	120	156.2	69	102	38.88
239	115	154.4	68	101	38.33
230	110	152.6	67	100.4	38
228.2	109	150.8	66	100	37.77
226.4	108	149	65	99.5	37.50
224.6	107	147.2	64	99	37.22
222.8	106	145.4	63	98.6	37
221	105	143.6	62	98	36 66
219.2	104	141.8	61	97	36.11
217.4	103	140	60	96.8	36
215.6	102	138.2	59	96	35.55
213.8	101	136.4	58	95	35
212	100	134.6	57	94	34.44
210.2	99	132.8	56	93.2	34
208.4	98	131	55	93	33.88

Table of Thermometric Equivalents.—*Continued.*

Fahrenheit.	Centigrade or Celsius.	Fahrenheit.	Centigrade or Celsius.	Fahrenheit.	Centigrade or Celsius.
92	33.33	48.2	9	—7.6	—22
91.4	33	46.4	8	—9.4	—23
91	32.77	44.6	7	—11.2	—24
90.5	32.50	42.8	6	—13	—25
90	32.22	41	5	—14.8	—26
89.6	32	39.2	4	—16.6	—27
89	31 66	37.4	3	—18.4	—28
88	31.11	35.6	2	—20.2	—29
87.8	31	33.8	1	—22	—30
87	30.55	32	0	—23.8	—31
86	30	30.2	— 1	—25.6	—32
85	29 44	28.4	— 2	—27.4	—33
84.2	29	26 6	— 3	—29.2	—34
84	28 88	24 8	— 4	—31	—35
83	28.33	23	— 5	—32.8	—36
82.4	28	21.2	— 6	—34 6	—37
82	27.77	19.4	— 7	—36.4	—38
81.5	27.50	17.6	— 8	—38.2	—39
81	27.22	15.8	— 9	—40	—40
80.6	27	14	—10		
80	26.66	12.2	—11		
79	26.11	10 4	—12		
78.8	26	8.6	—13		
78	25 55	6.8	—14		
77	25	5	—15		
76	24.44	3.2	—16		
75.2	24	1.4	—17		
75	23.88	0	—17.77		
74	23.33	—0.4	—18		
73.4	23	—2.2	—19		
71 6	22	—4	—20		
69.8	21	—5.8	—21		
68	20				
66.2	19				
64.4	18				
62.6	17				
62	16.66				
60.8	16				
60	15.55				
59	15				
57.2	14				
55.4	13				
53 6	12				
51.8	11				
50	10				

PART II.

METRIC
PRESCRIPTION FORMULARY.

SELECTED PRESCRIPTIONS

USED IN

HOSPITAL AND OUT-PATIENT PRACTICE.

A.

SOLUTIONS.*

No. 1.

℞. POTASS. ACET., ... 100 Gm
    AQUA, a sufficient quantity
      to make the finished solution
      measure . . . . . . 100 fGm
Dissolve and filter.
Each fGm represents 1 Gm of the salt.

No. 2.

℞. POTASS. IODIDUM, ... 100 Gm
    AQUA FERVIDA, a sufficient quantity
      to make . . . . . . 100 fGm
Dissolve and filter.
Each fGm represents 1 Gm of the salt.

No. 3.

℞. AMMON. BROMIDUM, ... 50 Gm
    AQUA, a sufficient quantity
      to make . . . . . . 100 fGm
Dissolve and filter.
Each fGm represents 50 cents of the salt.

---

* These solutions are convenient in dispensing, but should be fresh, and hence not kept ready made unless in constant demand. Solutions of salts containing alkaloids or organic acids are especially prone to undergo change, or become turbid and unsightly by the formation of confervoid matter.

No. 4.

R. Sod. Bromid., ... 50 Gm
Aqua, a sufficient quantity
    to make . . . . 100 fGm

Dissolve and filter.
Each fGm represents 50 cents of the salt.

No. 5.

R. Chloral Hydras, . . . 50 Gm
Aqua, sufficient
    to make . . . . . . 100 fGm.

Dissolve and filter.
Each fGm represents 50 cents of the salt.

No. 6.

R. Magnes. Sulph., ... 50 Gm
Aqua, sufficient
    to make . . . . . . 100 fGm

Dissolve and filter.
Each fGm represents 50 cents of the salt.

No. 7.*

R. Ferric. Pot. Tartr.,. . . . 50 Gm
Aqua Fervida, sufficient
    to make . . . . . . 100 fGm

Dissolve.
Each fGm represents 50 cents of the salt.

---

* Does not keep very long without getting mouldy.

No. 8.

R. POTASS. BICARB., ... 20 Gm
    AQUA, sufficient
      to make . . . . . . 100 fGm

Dissolve and filter.
Each fGm represents 20 cents of the salt.

No. 9.

R. POTASS. BROMID., ... 20 Gm
    AQUA, sufficient
      to make . . . . . . 100 fGm

Dissolve and filter.
Each fGm represents 20 cents of the salt.

No. 10.

R. AMMON. CHLORID., ... 20 Gm
    AQUA, sufficient
      to make . . . . . . 100 fGm

Dissolve and filter.
Each fGm represents 20 cents of the salt.

No. 11.

R. QUININ. SULPH., ... 20 Gm
    ACID. SULPH. DIL., sufficient.
    AQUA, a sufficient quantity
      to make the finished solution
      measure . . . . . . 100 fGm

Dissolve and filter.
Each fGm represents 20 cents of the salt.

No. 12.

℞. CINCHONID. SULPH., ... 20 Gm
ACID. SULPH. DIL., sufficient. '
AQUA, a sufficient quantity
to make the finished solution meas-
ure . . . . 100 fGm
Dissolve and filter.
Each fGm represents 20 cents of the salt.

No. 13.

℞. CINCHONIN. SULPH., ... 20 Gm
ACID. SULPH. DIL., sufficient.
AQUA, a sufficient quantity
to make the finished solution meas-
ure . . . . 100 fGm
Dissolve and filter.
Each fGm represents 20 cents of the salt.

No. 14.

℞. MORPH. SULPH., ... 10 Gm
AQUA, sufficient
to make . . . . . . 100 fGm
Dissolve and filter.
Each fGm represents 10 cents of the salt.

No. 15.

℞. MORPH. PURA, ... 10 Gm
ACID. CITR., ... 5 Gm
ALCOHOL, ... 10 fGm
AQUA, a sufficient quantity
to make the finished solution meas-
ure, ... 100 fGm
Triturate until dissolved and filter.
Each fGm represents 10 cents of Morphia.

No. 16.

R. Hydrarg. Chlorid. Corros., . . 5 Gm
Aqua Ebullient, sufficient
to make, ... 100 fGm

Dissolve and filter.
Each fGm represents 5 cents of the salt.

No. 17.

R. Potass. Chloras, ... 5 Gm
Aqua, sufficient
to make, ... 100 fGm

Dissolve and filter.
Each fGm represents 5 cents of the salt.

No. 18.*

R. Ant. Pot. Tartr., ... 2 Gm
Aqua, sufficient
to make, ... 100 fGm

Dissolve and filter.
Each fGm represents 2 cents of the salt.

No. 19.†

AQUA CHLOROFORMUM.

R. Chlorof. Purific., ... 2 fGm
Aqua, sufficient
to make, ... 100 fGm

Mix.

---

* Does not keep long without getting mouldy.

† A good vehicle for unpleasantly tasting remedies with which it may be properly combined. The small amount of Chloroform, which is itself not unpleasant, temporarily benumbs the gustatory nerve.

No. 20.

R. Acid. Carbol. Cryst., . . . 100 Gm
Aqua Fervida, ... 10 fGm

Warm the Carbolic Acid until liquefied. Then add the Water and mix thoroughly by agitation. Each fGm of the finished product represents very nearly 1 Gm of Phenol.

No. 21.

R. Acid. Carbol. Cryst., . . . 10 Gm
Glycerinum, sufficient
    to make, ... 100 fGm

Mix.

Each fGm represents 10 cents of Carbolic Acid.

No. 22.

R. Acid. Carbol. Cryst., . . . 1 Gm
Aqua, sufficient
    to make, ... 100 fGm

Dissolve.

Each fGm represents 1 cent of Carbolic Acid.

No. 23.*

R. Acid. Boric., ... 3 Gm
Aqua Fervida., sufficient
    to make, ... 100 fGm

Dissolve.

Each fGm represents 3 cents of Boric Acid.

No. 24.

R. Alumin. Pot. Sulph., . . . 2 Gm
Aqua, sufficient
    to make, ... 100 fGm

Dissolve and filter.

Each fGm represents 2 cents of Alum.

---

* For Lister's antiseptic treatment.

No. 25.

℞. ZINC. SULPH., ... 10 Gm
   AQUA, sufficient
     to make, ... .100 fGm
Dissolve and filter.
Each fGm represents 10 cents of the salt.

No. 26.

℞. ARGENT. NITR., ... 5 Gm
   AQUA DESTILL., sufficient
     to make, ... 100 fGm
Dissolve.
Each fGm represents 5 cents of the salt.

B.

INTERNAL REMEDIES.

MIXTURES.*

DIAPHORETICS, ETC.

No. 27.

℞. POTASS. NITR., ... 2.50 Gm
   ÆTHER. NITR. SPIR., . . . 7.50 fGm
   VINUM IPECAC., ... 2.50 fGm
   AQUA, a sufficient quantity
     to make the finished mixture meas·
     ure, ... 100 fGm
Mix. Dose, 5 to 10 fGm.

---

* Among the eligible vehicles with which to bring extemporaneous mixtures to the required volume, and at the same time remedy, as far as practicable, their unpleasant taste, may be mentioned Elix. Aurantii, Elix. Taraxaci, Syrups, Aromatic Waters, Fluid Extract of Liquorice (the latter only for mixtures in which there is no free acid), etc.

No. 28.

R. Vin. Ipecac, ...   5 fGm
  Tinct. Opium, ...   5 fGm
  Æther. Nitr. Spir.,   .   .   .   90 fGm
                                      100 fGm
  Mix. Dose, 5 fGm.

No. 29.

R. Tinct. Aconit. Rad.,   .   .   .   2 fGm
  Æther. Nitros. Spir.,   .   .   .   20 fGm
  Sol. Ammon. Acet.,   .   .   .   78 fGm
                                      100 fGm
  Mix.   Dose, 5 to 10 fGm.

No. 30.

R. Tinct. Aconit. Rad., ...50 fGm
  Æther. Nitros. Spir.,   .   .   .   50.00 fGm
  Liq. Potass. Citras,   .   .   .   49.50 fGm
                                      100 fGm
  Mix.   Dose, 20 fGm.

No. 31.

R. Potass. Nitr., ...   2.50 Gm
  Aqua Camphora, ...   30.00 fGm
  Æther. Nitros. Spir.,   .   .   .   20.00 fGm
  Sol. Ammon. Acet., sufficient
    to make the whole,   .   .   .   100 fGm
  Mix.   Dose, 10 fGm.

No. 32.

"*Imperial Drink.*"

R. Potass. Bitartr., ...   10 Gm
  Saccharum, ...   30 Gm
  Aqua, ...   1000 fGm
  Mix.

ANTI-RHEUMATICS.

No. 33.

℞. POTASS. IODID., ...	· 12.50 Gm
VIN. COLCH. SEM., ...	12.50 fGm
TINCT. GUAIACUM, ...	12.50 fGm
TINCT. GENT. COMP., . . .	25 fGm
TINCT. CINCH. COMP., sufficient	
to make, ...	100 fGm

Mix. Dose, 5 fGm.

No. 34.

℞. POTASS. NITR., ...	5 Gm
SOD. POT. TARTR., ...	5 Gm
VIN. COLCH. SEM., ...	10 fGm
AQUA, sufficient	
to make, ...	100 fGm

Mix. Dose, 5 fGm.

No. 35.

℞. MAGNES. SULPH., ...	10 Gm
MAGNES. CARB., ...	5 Gm
VIN. COLCH. SEM., ...	5 fGm
AQUA.MENTH. PIP., sufficient	
to make the whole, . . .	100 fGm

Mix. Dose, 20 fGm.

No. 36.

℞. POTASS. NITR., ...	2 Gm
VIN. COLCH. SEM., ...	20 fGm
EXTR. CIMICIF. FL., ...	20 fGm
SYRUPI, ...	30 fGm
AQUA, sufficient	
to make, ...	100 fGm

Mix. Dose, 5 fGm.

No. 37.

R. POTASS. IODID., ... 10 Gm
VIN. COLCH. SEM., ... 10 fGm
TINCT. CIMICIF., ... 10 fGm
AQUA, sufficient
   to make the whole, . . . 100 fGm
Mix. Dose, 10 fGm.

No. 38.

R. POTASS. IODID., ... 2 Gm
TINCT. COLCH. RAD., . . . 20 fGm
AQUA, sufficient
   to make the whole, . . . 100 fGm
Mix. Dose, 5 fGm.

SALICYLIC ACID MIXTURES.*

No. 39.

R. ACID. SALYCYL., ... 5 Gm
POTASS. ACET., ... 10 Gm
GLYCERINUM, ... 15 Gm
AQUA, sufficient
   to make the whole, . . . 100 fGm
Mix. Dose, 5 fGm.

No. 40.

R. ACID. SALICYL., ... 5 Gm
POTASS. IODID., ... 3 Gm
SYR. PRUN. VIRG., ... 20 fGm
SOL. AMMON. ACET., sufficient
   to make the whole, . . . 100 fGm
Mix. Dose, 5 fGm.

---

* Probably the least disagreeable Salicylic Acid mixture that can be made is one in which the acid is suspended (not dissolved) by thick syrup.

No. 41.

R. Sod. Salicyl., ... 15 Gm
   Tinct. Opii Deod., ... 7.50 fGm
   Syr. Simpl., ... 50 fGm
   Aqua, sufficient
      to make the whole, . . . 100 fGm
Mix. Dose, 5 fGm.

No. 42.

R. Acid. Salicyl., ... 2 Gm
   Ammon. Carb., ... 3 Gm
   Potass. Iodid., ... 4 Gm
   Chloral Hydr., ... 3 Gm
   Extr. Colch. Sem. Fl., . . . 1 fGm
   Aqua, sufficient
      to make the whole, . . . 100 fGm
Mix. Dose, 10 fGm.

ALTERATIVES, ETC.

No. 43.

R. Potass. Iodid., ... 5 Gm
   Inf. Gent. Comp., sufficient
      to make, ... 100 fGm
Mix. Dose, 5 fGm.

No. 44.

R. Potass. Iodid., ... 8 Gm
   Extr. Stilling. Fl., . . . 30 fGm
   Aqua, sufficient
      to make the whole, . . . 100 fGm
Mix. Dose, 5 fGm.

No. 45.

℞. POTASS. IODID., ...	10 Gm
AMMON. IODID., ...	5 Gm
AQUA, ...	20 fGm
TINCT. CINCH. COMP., sufficient	
to make the whole, . . .	100 fGm

Mix. Dose, 5 fGm.

No. 46.

℞. HYDRARG. IODID. RUBR., . . .	0.05 Gm
POTASS. IODID., ...	5.00 Gm
AQUA, sufficient	
to make the whole	100 fGm

Mix. Dose, 5 fGm.

No. 47.

℞. HYDRARG. CHLOR. CORROS., . .	0.10 Gm
POTASS. IODID., ...	10.00 Gm
TINCT. GENT. COMP., sufficient	
to make the whole	100 fGm

Mix. Dose, 5 fGm.

No. 48.

℞. HYDRARG. IODID. RUBR., . . .	0.10 Gm
POTASS. IODID., ...	7.50 Gm
SYR. SARSAP. COMP., ...	50.00 fGm
AQUA MENTH. PIP., sufficient	
to make the whole	100 fGm

Mix. Dose, 5 fGm.

No. 49.

R. Hydrarg. Iodid. Rubr., . . . 0.10 Gm
Ammon. Iodid., ... 5.00 Gm
Potass. Iodid., ... 10.00 Gm
Syr. Aurant. Cort., ... 30.00 fGm
Aqua, sufficient
  to make the whole . . . . 100 fGm
Mix. Dose, 5 fGm.

No. 50.

R. Liqu. Pot. Arsenit., ... 5 fGm
Elix. Aurant., sufficient
  to make the whole . . . . 100 fGm
Mix. Dose, 5 fGm.

No. 51.

R. Liqu. Arsen. Chlorid., . . . 5 fGm
Tinct. Ferr. Chlorid., . . . 5 fGm
Elix. Aurant., sufficient
  to make the whole . . . . 100 fGm
Mix. Dose, 5 fGm.

No. 52.

R. Arsen. Iodid., ... 0.04 Gm
Syr. Ferr. Iodid., ... 12.00 fGm
Potass. Iodid., ... 6.00 Gm
Elix. Aurant., sufficient
  to make the whole . . . . 100 fGm
Mix. Dose, 5 to 10 fGm.

6

No. 53.*

R. LIQU. ARSEN. CHLOR.,   .   .   .   8.00 fGm
TINCT. FERRIC. CHLOR.,   .   .   .   25.00 fGm
CINCHONID. SULPH., ...   4.00 Gm
STRYCHNIN. NITR., ...   0.05 Gm
ELIX. AURANT., sufficient
  to make the whole .   .   .   .   100 fGm

Mix. Dose, 5 fGm.

TONICS, ANTIPERIODICS, ETC.

No. 54.

R. ACID. HYDROCHLOR. DIL., .   .   .   3 fGm
INF. GENT. COMP., sufficient
  to make the whole .   .   .   .   100 fGm

Mix. Dose, 20 fGm.

No. 55.

R. ACID. NITROMUR. DIL.,   .   .   .   6.5 fGm
INF. GENT. COMP., sufficient
  to make the whole .   .   .   .   100 fGm

Mix. Dose, 10 fGm.

No. 56.

R. TINCT. FERRIC. CHLOR.,   .   .   .   20 fGm
SYRUPUS SACCHARUM, ...   20 fGm
GLYCERINUM, ...   30 fGm
AQUA, ...   30 fGm
                                        100 fGm

Mix. Dose, 5 fGm.

* Elixir Ferri Quiniæ et Strychniæ Phosphatum is an excellent preparation, to which Fowler's Solution could well be added.

No. 57.

R. Tinct. Ferric. Chlorid.,	. . .	3.50 fGm
Acid. Acet. Dil.,	. . .	3.50 fGm
Sol. Ammon. Acet.,	. . .	93.00 fGm
		100 fGm

Mix. Dose, 10 to 20 fGm.

No. 58.

R. Acid. Phosph. Dil.,	. . .	5 fGm
Tinct. Ferric. Chlor.,	. . .	10 fGm
Syr. Limon.,	. . . .	85 fGm
		100 fGm

Mix. Dose, 10 fGm.

No. 59.

R. Ferr. Sulph.,	. . . .	0.50 Gm
Potass. Carb.,	. . . .	0.60 Gm
Myrrh. Pulv.,	. . . .	1.50 Gm
Sacch. Pulv.,	. . . .	1.50 Gm
Spir. Lavend. Co.,	. . .	5.00 fGm
Aqua Cinnam., sufficient		
to make the whole .	. . .	100 fGm

Mix. Dose, 20 fGm.
(The Iron Sulphate should be added last, previously dissolved in about 5 fGm of the Cinnamon-water.)

No. 60.

R. Potass. Iodid.,	. . . .	10 Gm
Syr. Ferr. Iodid.,	. . .	20 fGm
Tinct. Calumba, sufficient		
to make the whole .	. . .	100 fGm

Mix. Dose, 5 fGm.

No. 61.

R. Ferric. Pot. Tartr.,	. . .	3 Gm
Tinct. Cinch. Comp., sufficient		
to make .	. . . .	100 fGm

Mix. Dose, 5 fGm.

No. 62.

R. QUIN. SULPH., ... 2.50 Gm
TINCT. FERRIC. CHLOR., . . . 10.00 fGm
AQUA, sufficient
to make the whole . . . . 100 fGm

Mix. Dose, 5 fGm.

No. 63.

R. QUIN. SULPH., ... 1 Gm
TINCT. FERRIC. CHLOR., . . . 10 fGm
SPIR. CHLOROFORM., ... 15 fGm
GLYCERIN., sufficient
to make the whole . . . ; 100 fGm

Mix. Dose, 5 fGm.

No. 64.

R. CINCHONIN. SULPH., ... 1 Gm
AQUA, ... 20 fGm
ACID. CITR., ... 1 Gm
SYR. SACCH., ... 30 fGm
TINCT. FERRIC. CHLOR., . . . 4 fGm
ELIX. AURANT., sufficient
to make the whole . . . . 100 fGm

Mix. Dose, 5 fGm.

No. 65.

R. CINCHONID. SULPH., ... 2 Gm
TINCT. FERRIC. CHLORID., . . . 10 fGm
SPIR. CHLOROF., ... 12 fGm
ELIX. AURANT., ... 78 fGm
100 fGm

Mix. Dose, 5 fGm.

No. 66.*

℞. QUININ. SULPH., ...	2 Gm
TINCT. FERRIC. CHLOR., . . .	10 fGm
GLYCERINUM, ...	5 fGm
ELIX. AURANT., ...	85 fGm
	100 fGm

Mix. Dose, 5 fGm.

No. 67.

℞. STRYCHNIN. NITR., ...	0.03 Gm
TINCT. CARD. COMP., ...	1 fGm
ALCOHOL, ...	5 fGm
AQUA, ...	5 fGm
SYRUP. SACCH., sufficient	
to make the whole	100 fGm

Mix. Dose, 5 fGm.

No. 68.

℞. TINCT. FERRIC. CHLOR., . . .	10 fGm
TINCT. NUX VOM., ...	10 fGm
ELIX. AURANT., ...	80 fGm
	100 fGm

Mix. Dose, 5 fGm.

No. 69.

℞. CINCHONIDIN. SULPH., . . .	1 Gm
ACID. SULPH. DIL., sufficient.	
ELIX. AURANT., sufficient	
to make the whole	100 fGm

Mix.

---

* Elixir Calisaya Ferratum is a much better preparation than such mixtures as Nos. 61 to 66, where therapeutically satisfactory. See p. 109.

No. 70.*

R. QUIN. SULPH., ... 1 Gm
ACID. SULPH. DIL., sufficient.
ELIX. AURANT., sufficient
    to make the whole . . . . 100 fGm
Mix.

No. 71.

R. CINCHONID. SULPH., ... 2 Gm
POTASS. IODID., ... 2 Gm
ACID. CITRIC., ... 2 Gm
AQUA DESTILL., sufficient
    to make the whole . . . . 100 fGm
Make a solution. Dose, 10 fGm.

          (Surgeon John Vansant, U. S. M. H. S.)

No. 72.

R. CINCHONID. SULPH., ... 2 Gm
POTASS. BROMID., ... 2 Gm
ACID. CITRIC., ... 2 Gm
AQUA, sufficient
    to make the whole . . . . 100 fGm
Make a solution. Dose, 10 fGm.

FOR DISEASES OF THE NERVOUS SYSTEM.

No. 73.

R. TINCT. CANNAB. IND., . . 12 fGm
SPIR. MENTH. PIP., . . . 1 fGm
GLYCERIN., sufficient
    to make the whole . . . 100 fGm
Mix. Dose, 5 fGm.

---

* The corresponding elixirs are much better preparations, unless the free acid is desired in the mixture.

No. 74.

℞. POTASS. BROMID.,	.	.	.	40 Gm
TINCT. BELLAD., .	.	.	.	8 ſGm
AQUA, sufficient				
to make the whole .	.	.	.	100 ſGm

Mix. Dose, 5 ſGm.

No. 75.

℞. POTASS. BROMID.,	.	.	.	50 Gm
EXT. CONIUM FL.,	.	.	. ·	25 ſGm
AQUA, sufficient				
to make the whole .	.	.	.	100 ſGm

Mix. Dose, 5 ſGm.

No. 76.

℞. SOD. BROMID.,	.	.	.	4 Gm
POTASS. BROMID.,	.	.	.	4 Gm
AMMON. BROMID.,	.	.	.	4 Gm
POTASS. IODID.,	.	.	.	2 Gm
AMMON. IODID., .	.	.	.	2 Gm
AMMON. CARB., .	.	.	.	1.30 Gm
TINCT. CALUMBA,	.	.	.	15 ſGm
AQUA, sufficient				
to make the whole .	.	.	.	100 ſGm

Mix. Dose, 10 to 15 ſGm.

No. 77.

℞. POTASS. BROMID.,	.	.	.	10 Gm
TINCT. VALER. AMMON.,	.	.	20 ſGm
TINCT. LUPULINUM,	.	.	.	30 ſGm
TINCT. DIGITALIS,	.	.	.	30 ſGm
AQUA, sufficient				
to make the whole .	.	.	.	100 ſGm

Mix. Dose, 20 ſGm.

No. 78.

R. POTASS. BROMID.,	. . . .	30 Gm
AMMON. BROMID.,	. . . .	30 Gm
EXTR. ERGOTA FL.,	. . . .	15 fGm
AQUA,	. . . . . .	25 fGm
		100 fGm

Mix. Dose, 10 fGm.

No. 79.

R. CHLORAL HYDR.,	. . . .	20 Gm
EXT. CONIUM SEM. FL.,	. . .	20 fGm
EXT. HYOSCYAM. FL., .	. . .	20 fGm
AQUA, sufficient		
to make the whole .	. . .	100 fGm

Mix. Dose, 5 fGm.

No. 80.

R. CHLORAL. HYDR.,	. . . .	4 Gm
SYR. AURANT. FLOR., .	. . .	10 fGm
SYR. TOLUT.,	. . . . .	10 fGm
AQUA CINNAM., sufficient		
to make the whole .	. . .	100 fGm

Mix.

No. 81.

R. CHLORAL HYDR.,	. . . .	4 Gm
POTASS. BROMID.,	. . . .	8 Gm
TINCT. OPIUM,	. . . . .	8 fGm
SYRUP TOLUT., .	. . . .	50 fGm
AQUA CINNAM., sufficient		
to make the whole .	. . .	100 fGm

Mix. Dose, 10 fGm.

No. 82.

R. Quinin. Sulph., ... 1.30 Gm
Ferric. Phosph., ... 1.30 Gm
Strychnina, ... 0.04 Gm
Acid. Phosph. Dil., a sufficiency.
Syrup. Zingiber, ... 35 fGm
Aqua, sufficient
to make . . . . . . 100 fGm

Mix. Dose, 5 fGm.

No. 83.

R. Oleum Phosphorat. (see p. 231), . 10 fGm
Vitell. Ovum, ... 5 Gm
Glycerinum, ... 5 fGm
Spir. Frumentum, sufficient
to make . . . . . . 100 fGm

Mix. Each fGm represents 1 mill of Phosphorus.

FOR DISEASES OF THE RESPIRATORY SYSTEM.

No. 84.

R. Syr. Scilla, ... 25 fGm
Syr. Ipecac., ... 25 fGm
Tinct. Opium Camph., . . . 25 fGm
Glycerinum, ... 25 fGm
 100 fGm

Mix. Dose, 5 fGm.

No. 85.

R. Syr. Tolut., ... 30 fGm
Syr. Ipecac., ... 30 fGm
Muc. Acacia, ... 20 fGm
Tinct. Opium Camph., . . . 10 fGm
Tinct. Lobelia, ... 10 fGm
 100 fGm

Mix. Dose, 5 fGm.

No. 86.

℞. Ant. Pot. Tartr.,	. . . .	0.05 Gm
Morph. Sulph.,	. . . .	0.06 Gm
Aqua,	. . . . . .	50 fGm
Syr. Tolut.,	. . . . .	50 fGm
		100 fGm

Mix. Dose, 5 fGm.

No. 87.

℞. Morph. Sulph.,	. . . .	0.05 Gm
Ant. Pot. Tartr.,	. . . .	0.05 Gm
Ammon. Chlorid.,	. . . .	10 Gm
Syr. Tolut.,	. . . . .	50 fGm
Aqua,		
sufficient to make the whole	.	100 fGm

Mix. Dose, 5 fGm.

No. 88.

℞. Ammon. Chlorid.,	. . . .	5 Gm
Syr. Senega,	. . . . .	15 fGm
Syr. Scilla,	. . . . .	15 fGm
Syr. Prun. Virg.,	. . .	50 fGm
Morph. Sulph.,	. . . .	0.06 Gm
Extr. Glycyrrh. Fl.,	. . .	20 fGm
		100 fGm

Mix. Dose, 5 fGm.

No. 89.

℞. Tinct. Sanguinar.,	. . .	8 fGm
Tinct. Opium Camph.,	. . .	10 fGm
Syr. Scilla,	. . . . .	10 fGm
Syr. Tolut.,	. . . . .	10 fGm
Aqua,	. . . . . .	62 fGm
		100 fGm

Mix. Dose, 5 fGm.

No. 90.

R. Tinct. Opium Camph., . . . 10 fGm
    Spir. Ammon Arom., . . . 15 fGm
    Syr. Prun. Virg., ... 20 fGm
    Extr. Ipecac. Fl., ... 1.50 fGm
    Extr. Glycyrrh. Fl., . . . 5 fGm
    Aqua, sufficient
      to make the whole . . 100 fGm

Mix. Dose, 5 to 10 fGm.

No. 91.

R. Morph. Acet., ... 0.10 Gm
    Acid. Acet. Dil., ... 5 fGm
    Syr. Prun. Virg., ... 30 fGm
    Syr. Ipecac., ... 30 fGm
    Syr. Tolut., ... 30 fGm
    Aqua, sufficient
      to make the whole . . . 100 fGm

Mix. Dose, 5 fGm.

No. 92.

R. Morph. Sulph., ... 0.05 Gm
    Syr. Scilla, ... 25 fGm
    Syr. Ipecac., ... 25 fGm
    Syr. Tolut., ... 20 fGm
    Syr. Prun. Virg., ... 20 fGm
    Tinct. Benz. Comp., . . . 5 fGm
    Tinct. Sanguinar., ... 5 fGm
                     100 fGm

Mix. Dose, 5 fGm

No. 93.

℞. Ammon. Chlorid.,	. . . .	2 Gm
Potass. Chlor.,	. . . .	3 Gm
Aqua,	. . . . . .	50 fGm
Syr. Senega,	. . . . .	10 fGm
Syr. Ipecac.,	. . . . .	10 fGm
Syr. Tolut.,	. . . . .	25 fGm
Extr. Glycyrrh. Fl.,	. . .	3 fGm
Tinct. Cubeba,	. . . .	2 fGm
		100 fGm

Mix. Dose, 5 fGm.

No. 94.

℞. Ammon. Chlorid.,	. . . .	3.50 Gm
Extr. Glycyrrh. Fl.,	. . .	3 fGm
Aqua Fœnicul.,	. . . .	97 fGm
		100 fGm

Mix. Dose, 20 fGm.

No. 95.

℞. Ammon. Chlorid.,	. . . .	3 Gm
Spir. Æther. Comp.,	. . . .	20 fGm
Syr. Prun. Virg.,	. . . .	60 fGm
Aqua, sufficient		
to make the whole .	. . .	100 fGm

Mix. Dose, 5 fGm.

No. 96.

(STOKES'S EXPECTORANT.)

℞. Ammon. Carb.,	. . . . .	1.50 Gm
Ext. Scilla Fl.,	. . . .	3 fGm
Ext. Senega Fl.,	. . .	3 fGm

TINCT. OPIUM CAMPH., . . . 20 fGm
SYR. TOLUT., ... 20 fGm
AQUA, sufficient
   to make the whole . . . . 100 fGm
Mix. Dose, 5 fGm.

No. 97.

(BROWN MIXTURE.)

℞. EXTR. GLYCYRRH. PULV., . . . 3 Gm
   SACCH. PULV., ... 3 Gm
   ACACIA PULV., ... 3 Gm
   TINCT. OPIUM CAMPH., . . . 12 fGm
   VIN. ANTIMON., ... 6 fGm
   SPIR. ÆTHER. NITROS., . . . 3 fGm
   AQUA, sufficient
     to make the whole . . . . 100 fGm
Mix. Dose, 10 to 20 fGm.

No. 98.

℞. EXT. IPECAC. FL., ... 0.60 fGm
   SPIR. ÆTHER. COMP., . . . 3.50 fGm
   MIST. GLYCYRRH. COMP. (No. 97), suffi-
     cient to make the whole . . . 100 fGm
Mix. Dose, 10 fGm.

No. 99.

℞. SPIR. ÆTHER. COMP.,. . . . 20 fGm
   TINCT. HYOSCYAM., ... 20 fGm
   SYR. PRUN. VIRG., ... 20 fGm
   SYR. TOLUT., ... 20 fGm
   AQUA, ... 20 fGm
                        100 fGm
Mix. Dose, 5 fGm.

No. 100.

R. Tinct. Lobelia, ... 15 fGm
Tinct. Hyoscyam., ... 25 fGm
Spir. Æther. Comp., . . . 25 fGm
Syrup. Tolut., ... 35 fGm
                                    100 fGm
Mix. Dose, 5 fGm.

No. 101.

R. Morph. Acet., ... 0.04 Gm
Tinct. Hyoscyam., ... 20 fGm
Syr. Tolut., ... 30 fGm
Aqua, ... 50 fGm
                                    100 fGm
Mix. Dose, 5 to 10 fGm.

No. 102.

R. Acid. Hydrocyan. Dil., . . . 2.50 fGm
~ Chloroform. Purific., . . . 2.50 fGm
Tinct. Hyoscyam., ... 25 fGm
Syrup. Tolut., ... 25 fGm
Aqua Camph., ... 25 fGm
Mucil. Acacia, ... 20 fGm
                                    100 fGm
Mix. Dose, 5 fGm.

No. 103.

R. Potass. Cyanid., ... 0.10 Gm
Morph. Acet., ... 0.05 Gm
Acet. Sanguin., ... 5 fGm
Syrup. Tolut., ... 50 fGm
Aqua, ... 45 fGm
                                    100 fGm
Mix. Dose, 5 fGm.

No. 104.

℞. SPIR. ÆTHER. COMP.,	.	.	25 fGm
SYR. IPECAC.,	. . .	.	25 fGm
TINCT. OPIUM CAMPH.,	.	.	25 fGm
AQUA,	. . .	.	25 fGm
			100 fGm

Mix. Dose, 5 fGm.

No. 105.

℞. SPIR. ÆTHER. COMP.,	.	.	5 fGm
SPIR. ÆTHER. NITROS.,	.	.	10 fGm
TINCT. OPIUM CAMPH.,	.	.	25 fGm
SYR. PRUN. VIRG.,	. .	.	30 fGm
SYR. SENEGA,	. . .	.	30 fGm
			100 fGm

Mix. Dose, 5 fGm.

No. 106.

℞. POTASS. IODID.,	. .	.	8 Gm
SYRUP. SACCH., .	. .	.	40 fGm
SPIR. ÆTHER. COMP.,	.	.	30 fGm
TINCT. TOLUT.,	. .	.	3 fGm
EXTR. PRUN. VIRG. FL.,	.	.	3 fGm
AQUA, sufficient			
to make the whole,			100 fGm

Mix. Dose, 5 fGm.

No. 107.

℞. LIQU. MORPH. SULPH. (U. S. P.),	.	50 fGm
SPIR. ÆTHER. COMP.,	. . .	50 fGm
		100 fGm

Mix. Dose, 5 fGm. To be well shaken.

No. 108.

R. Tinct. Opium, ... 10 fGm
Tinct. Stramonium, ... 10 fGm
Tinct. Lobelia, ... 20 fGm
Æther. Fort., ... 60 fGm
----
100 fGm

Mix. Dose, 5 fGm.

No. 109.

R. Acid. Nit. Dil., ... 4 fGm
Syr. Prun. Virg., ... 16 fGm
Aqua, ... 80 fGm
----
100 fGm

Mix. Dose, 5 fGm.

No. 110.

R. Acid. Hydrobrom. (34 per cent.), . 10 fGm
Spir. Chloroformum, ... 10 fGm
Syr. Scilla, ... 25 fGm
Aqua, ... 55 fGm
----
100 fGm

Mix. Dose, 10 fGm.

No. 111.

R. Morph. Sulph., ... 0.08 Gm
Chloral. Hydrat., ... 7.50 Gm
Potass. Bromid., ... 7.50 Gm
Syr. Scilla, ... 15 fGm
Syr. Ipecac., ... 15 fGm
Syr. Prun. Virg., sufficient
to make the whole . . . . 100 fGm

Mix. Dose, 5 fGm.

No. 112.

R. Quin. Sulph., ...   2 Gm
    Potass. Bromid., ...   6 Gm
    Elix. Aurant., ...   30 fGm
    Syr. Prun. Virg.,   . . .   30 fGm
    Extr. Glycyrrh. Fl.,   . .   5 fGm
    Aqua, sufficient
        to make the whole . . . .   100 fGm

Mix. Dose, 5 fGm.

No. 113.

R. Potass. Chlor., ...   2.50 Gm
    Tinct. Ferric. Chlor.,   . .   10 fGm
    Syr. Zingib., ...   15 fGm
    Aqua, sufficient
        to make the whole . . . .   100 fGm

Mix. Dose, 10 fGm.

No. 114.

R. Quin. Sulph., ...   1 Gm
    Potass. Chlor., ...   10 Gm
    Tinct. Ferric. Chlor.,   . .   15 fGm
    Glycerinum, ...   50 fGm
    Syrup. Sacch., sufficient
        to make the whole . . .   100 fGm

Mix. Dose, 5 to 10 fGm.

No. 115.

R. Creasotum, ...   1 fGm
    Mucil. Acacia, ...   50 fGm
    Syr. Sacch., ...   49 fGm
                         100 fGm

Mix. (Used in bronchitis.)

No. 116.

R. OLEUM MORRHUA, ... 60 fGm
LIQU. CALX, ... 20 fGm
SYR. PRUN. VIRG., ... 20 fGm
                                    100 fGm
Mix.   Dose, 20 fGm.

No. 117.

R. CALCIC. PHOSPH. PRÆCIP., . . . 10 Gm
OL. MORRHUA, ... 50 fGm
GLYCERINUM, ... 10 fGm
MUCIL. ACACIA, ... 40 fGm
ÆTHEROL AMYGD. AMAR., . . . 1 drop
                                    100 fGm
Mix.

No. 118.

R. ACID. GALLICUM, ... 1.50 Gm
MAGNES. SULPH., ... 10 Gm
ACID. SULPH. DIL., ... 1 fGm
EXTR. ERGOTA FL., ... 2.50 fGm
AQUA CINNAM., sufficient
   to make the whole . . . . 100 fGm
Mix.   Dose, 10 to 20 fGm.

FOR DISEASES OF THE DIGESTIVE SYSTEM.

No. 119.

R. PEPSIN. SACCHAR., ... 4 Gm
ACID. HYDROCHLOR. DIL., . . . 2 fGm
GLYCERINUM, ... 20 fGm
AQUA, sufficient
   to make the whole . . . . 100 fGm

Mix well, and, after standing an hour, filter.   Dose, 5 to 10 fGm.

No. 120.

R. SODIC. PHOSPH., ... 7.50 Gm
ACID. HYDROCHLOR. DIL., . . . 7.50 fGm
VIN. PEPSIN., ... 50 fGm
ELIX. AURANT., sufficient
to make the whole . . . . 100 fGm

Mix. Dose, 10 fGm.

No. 121.

R. BISMUTH. SUBNITR., ... 7.50 Gm
ACID. HYDROCYAN. DIL., . . . 1.50 fGm
TINCT. GENT. COMP., . . . 20 fGm
SYRUP. ZINGIB., ... 30 fGm
SPIR. AMMON. AROM., ... 7.50 fGm
AQUA, sufficient
to make the whole . . . . 100 fGm

Mix. Dose, 10 fGm. To be well shaken.

No. 122.

R. CREASOT., ... 1.50 fGm
ACID. HYDROCYAN. DIL., . . . 3 fGm
ACACIA PULV., ... 30 Gm
SACCH. PULV., ... 30 Gm
AQUA, sufficient
to make the whole . . . . 100 fGm

Mix. Dose, 5 fGm. (For children.)

No. 123.

R. SODIC. BICARB., ... 3 Gm
AQUA CREASOT., ... 25 fGm
MIST. CRETA, ... 75 fGm .
100 fGm

Mix. Dose, 20 fGm.

No. 124.

R. Sod. Bicarb., ...   15 Gm
Tinct. Zingib., ...   5 fGm
Tinct. Gent. Comp., ...   25 fGm
Aqua, sufficient
to make the whole .   .   .   .   100 fGm

Mix.   Dose, 10 fGm.

No. 125.

R. Sodic. Bicarb., ...   3 Gm
Inf. Gent. Comp., sufficient
to make .   .   .   .   .   100 fGm

Mix.   Dose, 20 fGm.

No. 126.

R. Sodic. Bicarb., ...   3 Gm
Extr. Rheum Fl.,   .   .   .   3 fGm
Spir. Menth. Pip.,   .   .   :   .   3 fGm
Aqua, sufficient
to make the whole .   .   .   .   100 fGm

Mix.   Dose, 5 fGm.

No. 127.

R. Bismuth. Subnitr., .   .   .   2 Gm
Potass. Bromid.,   .   .   .   2 Gm
Acid. Hydrocyan. Dil., .   .   .   0.60 fGm
Spir. Chlorof., ...   2 fGm
Mucil. Acacia, ...   30 fGm
Aqua, sufficient
to make the whole .   .   .   .   100 fGm

Mix.   Dose, 30 to 40 fGm.   To be well shaken.

No. 128.

℞. Oleum Ricinus, ... 10 fGm
   Mucil. Acacia, ... 10 fGm
   Tinct. Opium, ... 5 fGm
   Tinct. Rheum Arom., . . . 10 fGm
   Aqua Menth. Pip., sufficient
     to make the whole . . . . 100 fGm
Mix. Dose, 5 fGm. (For children.)

No. 129.

℞. Extr. Rheum Fl., ... 1.50 fGm
   Extr. Ipecac. Fl., ... 0.30 fGm
   Sodic. Bicarb., ... 3 Gm
   Glycerinum, ... 40 fGm
   Aqua Menth. Pip., sufficient
     to make the whole . . . . 100 fGm
Mix. Dose, 3 to 5 fGm. (For children.)

No. 130.

℞. Morph. Sulph., ... 0.10 Gm
   Chlorof. Purific., ... 15 fGm
   Syr. Zingiber., ... 85 fGm
                           100 fGm
Mix. Dose, 5 fGm.

No. 131.

℞. Tinct. Opium, ... 10 fGm
   Tinct. Capsicum, ... 10 fGm
   Spir. Camphora, ... 10 fGm
   Spir. Menth. Pip., ... 10 fGm
   Syr. Sacch., ... 20 fGm
   Aqua, ... 40 fGm
                           100 fGm
Mix. Dose, 3 to 5 fGm. (For children.)

No. 132.

R. TINCT. OPIUM, ... 20 fGm
TINCT. CAPSICUM, ... 20 fGm
TINCT. RHEUM AROM., . . . 20 fGm
SPIR. MENTH. PIP., ... 20 fGm
SPIR. CAMPHORA, ... 20 fGm
 ‾‾‾‾‾‾
 100 fGm.
Mix. Dose, 1 to 3 fGm.

No. 133.

R. TINCT. OPIUM, ... 20 fGm
TINCT. CAPSICUM, ... 20 fGm
TINCT. CAMPHORA, ... 20 fGm
CHLOROFORM. PURIFIC., . . . 20 fGm
ALCOHOL. FORT., ... 30 fGm
 ‾‾‾‾‾‾
 100 fGm
Mix. Dose, 1 to 5 fGm.

No. 134.

R. MORPH. SULPH., ... 0.10 Gm
ÆTHEROL. MENTH. PIP., . . . 0.10 fGm
ÆTHER. FORT., ... 2.50 fGm
ALCOHOL. FORT., ... 2.50 fGm
ACID. HYDROCYAN. DIL., . . . 5 fGm
CHLOROFORM. PURIF., ... 15 fGm
SYRUP. SACCH., sufficient
 to make the whole . . . . 100 fGm
Mix. (Substitute for so-called chlorodyne.)

No. 135.

R. TINCT. OPIUM CAMPH., . . . 15 fGm
TINCT. CATECHU, ... 15 fGm
MIST. CRETA, ... 70 fGm
 ‾‾‾‾‾‾
 100 fGm
Mix. Dose, 5 fGm. To be well shaken.

No. 136.

R. TINCT. OPIUM CAMPH.,	.	.	.	25 fGm
TINCT. KRAMERIA,	.	.	.	25 fGm
SPIR. LAVEND. COMP.,		.	.	4 fGm
MIST. CRETA,	.	.	.	46 fGm
				100 fGm

Mix. Dose, 10 to 20 fGm. To be well shaken.

No. 137.

R. TINCT. CAPSICUM,	.	.	.	5 fGm
TINCT. CATECHU,	.	.	.	15 fGm
TINCT. KINO,	.	.	.	15 fGm
TINCT. KRAMER.,	.	.	.	15 fGm
TINCT. OPIUM,	.	.	.	10 fGm
SPIR. MENTH. PIP.,	.	.	.	10 fGm
SPIR. CAMPH.,	.	.	.	15 fGm.
GLYCERINUM,	.	.	.	15 fGm
				100 fGm

Mix. Dose, 2 to 5 fGm.

No. 138.

R. ACID. SULPH. AROM., .	.	.	7.50 fGm
EXTR. HÆMATOXYL. FL.,	.	.	12.50 fGm
TINCT. OPIUM CAMPH.,	.	.	40 fGm
SYR. ZINGIBER, .	.	.	40 fGm
			100 fGm

Mix. Dose, 5 to 10 fGm.

No. 139.

R. ACID. SULPH. AROM.,	.	.	.	5 fGm
TINCT. OPIUM DEOD.,	.	.	.	5 fGm
SYRUP. RUBUS VILL., sufficient				
to make the whole	.	.	.	100 fGm

Mix. Dose, 20 fGm.

No. 140.

R. TINCT. OPIUM, ... 2.50 fGm
ACID. SULPH. AROM., . . . 2.50 fGm
AQUA CAMPH., ... 60 fGm
AQUA, sufficient
to make the whole . . . 100 fGm
Mix. Dose, 20 fGm.

No. 141.

R. ACID. NITR. FUM., . . . 1.50 fGm
TINCT. OPIUM, ... 1 fGm
AQUA CAMPH., sufficient
to make the whole . . . 100 fGm
Mix. Dose, 5 to 10 fGm.

No. 142.

R. ACID. SULPH. DIL., ... 4 fGm
SYRUP. SACCH., ... 16 fGm
AQUA, ... 80 fGm
100 fGm
Mix.

No. 143.

R. MAGNES. SULPH., ... 20 Gm
INF. SENNA, ... 80 fGm
SYR. ZINGIB., sufficient
to make the whole . . . 100 fGm
Mix. Dose, 40 fGm.

No. 144.

R. MAGNES. SULPH., ... 20 Gm
MAGNES. CARB., ... 5 Gm
AQUA MENTHA PIP., sufficient
to make the whole . . . 100 fGm
Mix. Dose, 40 to 50 fGm. To be well shaken.

No. 145.

℞. Potass. Sod. Tartr., . . . 10 Gm
Magnes. Sulph., ... 10 Gm
Aqua Mentha Pip., sufficient
to make the whole . . . 100 fGm

Mix. Dose, 20 to 40 fGm.

No. 146.

℞. Magnes. Sulph., ... 10 Gm
Magnes. Carb., ... 3 Gm
Rheum Pulv., ... 5 Gm
Aqua Mentha Pip., sufficient
to make the whole . . . 100 fGm

Mix. Dose, 20 fGm. To be well shaken.

No. 147.

℞. Sodic. Pot. Tartr., ... 50 Gm
Extr. Senna Fl., ... 5 fGm
Extr. Rheum Fl., ... 1.50 fGm
Extr. Gent. Fl., ... 1.50 fGm
Aqua, sufficient
to make the whole . . . 100 fGm

Mix.

No. 148.

℞. Resina Podophyllum, . . . 0.05 Gm
Sodic. Bicarb., ... 4 Gm
Extr. Rheum Fl., ... 4 fGm
Aqua Camph., ... 50 fGm
Inf. Gent. Comp., sufficient
to make the whole . . . 100 fGm

Mix. Dose, 20 fGm. To be well shaken before taken

No. 149.

R. Magnes. Sulph., ...   50 Gm
Extr. Senna Fl., ...   5 fGm
Ferr. Sulph. Exsicc.,   .   .   .   0.25 Gm
Acid. Sulph. Dil., ...   1 fGm
Aqua Mentha Pip., sufficient
to make the whole   .   .   .   100 fGm

Make a solution and filter.

No. 150.

R. Oleum Ricinus, ...   20 fGm
Acacia Pulv., ...   10 Gm
Saccharum Pulv., ...   10 Gm
Ætherol. Menth. Pip.,   .   .   .   2 drops
Aqua, sufficient
to make the whole   .   .   .   100 fGm

Mix.

No. 151.

R. Ætherol. Terebinth.,   .   .   .   5 fGm
Syr. Sacch., ...   25 fGm
Mucil. Acacia, ...   35 fGm
Aqua, ...   35 fGm
100 fGm

Mix.

FOR DISEASES OF THE URINARY SYSTEM, ETC.

No. 152.

R. Potass. Acet., ...   10 Gm
Spir. Æther. Nitr., ...   20 fGm
Aqua, sufficient
to make the whole   .   .   .   100 fGm

Mix. Dose, 10 to 20 fGm.

No. 153.

℞. Potass. Acet., ...	5 Gm
Acet. Scilla, ...	5 fGm
Decoct. Scopar., sufficient	
to make the whole	100 fGm

Mix. Dose, 20 fGm.

No. 154.

℞. Tinct. Digitalis, ...	3 fGm
Vin. Antimon., ...	10 fGm
Spir. Æther. Nitros., . . .	10 fGm
Syr. Acid. Citr., sufficient	
to make the whole	100 fGm

• Mix. Dose, 5 fGm.

No. 155.

℞. Tinct. Digitalis, ...	6 fGm
Spir. Æther. Nitr., ...	12 fGm
Potass. Iodid., ...	3 Gm
Potass. Acet., ...	6 Gm
Syr. Scilla, ...	12 fGm
Aqua, sufficient	
to make the whole	100 fGm

Mix. Dose, 5 to 10 fGm.

No. 156.

℞. Tinct. Digitalis, ...	1.50 fGm
Potass. Acet., ...	6 Gm
Spir. Æther. Nitr., . . .	25 fGm
Aqua, sufficient	
to make the whole	100 fGm

Mix. Dose, 20 fGm.

No. 157.

℞. Extr. Buchu Fl., ...	20 fGm
Spir. Æther. Nitros., . . .	20 fGm
Tinct. Opium Camph., . . .	20 fGm
Extr. Pareira Brav. Fl., . . .	20 fGm
Aqua Camphora, ...	20 fGm
	100 fGm

Mix. Dose, 5 fGm.

No. 158.

℞. Ant. Pot. Tartr., ...	0.02 Gm
Magnes. Sulph., ...	20 Gm
Succus. Limon., ...	30 fGm
Aqua, sufficient	
to make the whole	100 fGm ·

Mix. Dose, 20 fGm.

No. 159.

℞. Copaiba, ...	20 fGm
Liqu. Potassa, ...	5 fGm
Spir. Æther. Nitros., . . .	30 fGm
Mucil. Acacia, ...	15 fGm
Spir. Lavend. Comp., ...	15 fGm
Syr. Sacch., ...	15 fGm
	100 fGm

Mix. Dose, 10 fGm.

No. 160.

℞. Copaiba, ...	15 fGm
Liqu. Potassa, ...	5 fGm
Spir. Æther. Nitr., ...	15 fGm
Ætherol. Gaultheria, . . .	0.20 fGm
Extr. Glycyrrh. Fl., . . .	2 fGm
Syr. Acacia, sufficient	
to make the whole	100 fGm

Mix. Dose, 10 to 20 fGm.

No. 161.

℞. Copaiba, ...	10 fGm
Vitell. Ovum, ...	5 fGm
Syr. Sacch., ...	10 fGm
Aqua Menth. Pip., ...	75 fGm
	100 fGm

Mix.   Dose, 20 fGm.

No. 162.

℞. Copaiba, ...	10 fGm
Tinct. Ferr. Chlor., . . .	10 fGm
Tinct. Cantharid., ...	10 fGm
Glycerinum, ...	20 fGm
Syrup. Sacch., ...	50 fGm
	100 fGm

Mix.   Dose, 5 fGm.

No. 163.

℞. Tinct. Ferr. Chlor., . . .	1 fGm
Spir. Chloroform., ...	1 fGm
Tinct. Ergota, ...	2 fGm
Inf. Quassia, ...	96 fGm
	100 fGm

Mix.   Dose, 30 to 40 fGm.

POWDERS.*

No. 164.

R. Morph. Sulph., ...   10 Gm
Sacch. Lact. Pulv., ...   90 Gm
                                     ——————
                                     100 Gm

Mix.   Each gram contains 10 cents of the Morphina salt.

No. 165.

R. Camph. Pulv., ...   15 Gm
Pulv. Ipecac. Comp., ...   25 Gm
Potass. Nitras Pulv.,   .   .   .   60 Gm
                                     ——————
                                     100 Gm
Mix.   Dose, 1 Gm.

Nö. 166.

R. Ant. Pot. Tartr., ...   3 Gm
Ipecacuanha Pulv.,   .   .   .   97 Gm
                                     ——————
                                     100 Gm
Mix.   Dose, 1 Gm.

No. 167.

R. Morph. Sulph., ...   6 Gm
Ipecac. Pulv., ...   40 Gm
Resina Podophyll.,   .   .   .   4 Gm
Resina Leptandra,   .   .   .   10 Gm
Sod. Carb. Exsicc., ...   40 Gm
                                     ——————
                                     100 Gm
Mix.   Dose, 1 Gm.

---

* Milk Sugar is preferable to Cane Sugar as a diluent of powders, because of its hardness as well as its greater specific gravity and less solubility. It aids trituration, and does not, like Cane Sugar, dissolve rapidly, leaving heavy or insoluble substances, like Calomel, in the bottom of the spoon when administered.

No. 168.

℞. Senna Pulv., ...	20 Gm
Potass. Bitartr. Pulv., . . .	20 Gm
Sulph. Lotum, ...	20 Gm
Zingib. Pulv., ...	5 Gm
Sacch. Alb., ...	35 Gm
	100 Gm

Mix. Dose, 10 Gm.

No. 169.

℞. Magnes. Carb., ...	30 Gm
Sulph. Præcip., ...	40 Gm
Sod. Bicarb., ...	15 Gm
Sacch. Alb., ...	10 Gm
Zingiber, ...	5 Gm
	100 Gm

Mix. Dose, 5 Gm.

No. 170.

℞. Magnes. Sulph., ...	20 Gm
Magnes. Carb., ...	20 Gm
Sulphur, ...	20 Gm
Sacch. Alb., ...	25 Gm
Anisum Pulv., ...	15 Gm
	100 Gm

Mix. Dose, 5 Gm.

No. 171.

℞. Jalapa Pulv., ...	20 Gm
Potass. Bitartr. Pulv., . . .	80 Gm
	100 Gm

Mix. Dose, 5 Gm.

No. 172.

℞. HYDRARG. CHLOR. MIT., . . .	25 Gm
JALAPA PULV., ...	75 Gm
	100 Gm

Mix. Dose, 1 to 2 Gm.

No. 173.*

℞. ELATERIN. ALB., ...	10 Gm
SACCH. LACT., ...	90 Gm
	100 Gm

Mix. Dose, 4 Cents.

No. 174.

℞. BISMUTH. SUBNITR., ...	50 Gm
PEPSIN SACCHARAT., ...	50 Gm
	100 Gm

Mix. Dose, 1 to 2 Gm.

No. 175.

℞. MORPH. SULPH., ...	3 Gm
BISMUTH. SUBNITR., ...	97 Gm
	100 Gm

Mix. Dose, 50 Cents.

No. 176.

℞. BISMUTH. SUBNITR., ...	50 Gm
SOD. BICARB., ...	50 Gm
	100 Gm

Mix. Dose, 0.50 to 1 Gm.

---

* Other powerful remedies, such as the narcotic extracts, may be diluted the same way. See Report of American Pharmaceutical Association on Revision of Pharmacopœia, by Charles Rice, p. 38.

No. 177.

R. Bismuth. Subnitr., ... 80 Gm
Pulv. Ipecac. Comp., ... 20 Gm
_____
100 Gm

Mix. Dose 0.50 to 1 Gm.

No. 178.

R. Bismuth. Subnitr., ... 65 Gm
Pulv. Aromat., ... 35 Gm
_____
100 Gm

Mix. Dose, 1 Gm.

No. 179.

R. Bismuth. Subnitr., ... 50 Gm
Rheum Pulv., ... 25 Gm
Pulv. Aromat., . . . 25 Gm
_____
100 Gm

Mix. Dose, 1 Gm.

No. 180.

R. Pulv. Ipecac. Comp., . . . 20 Gm
Acid. Tannic., ... 20 Gm
Bismuth. Subnitr., ... 60 Gm
_____
100 Gm

Mix. Dose, 1 Gm.

No. 181.

R. Acid. Tannic., ... 30 Gm
Bismuth. Subnitr., ... 70 Gm
_____
100 Gm

Mix. Dose, 1 Gm.

No. 182.

R. Sod. Bicarb., ... 25 Gm
Bismuth. Subcarb., ... 50 Gm
Carb. Lignum. Pulv., . . . 15 Gm
Zingiber Pulv., ... 10 Gm
_____
100 Gm

Mix. Dose, 1 Gm.

No. 183.

R. RHEUM PULV., ...	20 Gm
MAGNES. OXID., ...	70 Gm
ZINGIBER. PULV., ...	10 Gm
	100 Gm

Mix.

No. 184.

R. RHEUM PULV., ...	40 Gm
SOD. BICARB., ...	40 Gm
ZINGIBER. PULV., ...	10 Gm
SACCH. ALB., ...	10 Gm
	100 Gm

Mix. Dose, 1 to 2 Gm.

No. 185.

R. HYDRARG. CHLOR. CORROS., . .	5 Gm
SOD. CARB. EXSICC., ...	30 Gm
PULV. CRETA COMP., ...	65 Gm
	100 Gm .

Mix. Dose, 20 to 50 Cents.

PILLS.*
No. 186.

R. ASAFŒTIDA., ...	5 Gm
CAMPHORA., ...	5 Gm
OPIUM, ...	1 Gm
EXTR. BELLAD., ...	1.50 Gm

Make 100 pills.

---

* Powdered tragacanth and glycerin—one, or the other, or both, as the case may require, but not previously mixed—are the best excipients for general use in forming good pill masses (sufficiently firm, and not liable to become hard and difficult to dissolve). Lycopodium is undoubtedly the best conspergative—to keep the pills from adhering together.

No. 187.

℞. Extr. Hyoscyam., ...	4 Gm
Extr. Conium, ...	4 Gm
Extr. Ignatia, ...	3 Gm
Extr. Opium, ...	3 Gm
Extr. Aconitum, ...	2 Gm
Extr. Cannab. Ind., ...	1.50 Gm
Extr. Stramonium, ...	1.20 Gm
Extr. Bellad., ...	1 Gm

Make 100 pills.

No. 188.

℞. Zinc. Oxid., ...	10 Gm
Extr. Hyoscyam., ...	10 Gm

Make 100 pills.

No. 189.

℞. Opium, ...	5 Gm
Camphora, ...	10 Gm

Make 100 pills.

No. 190.

℞. Opium Pulv., ...	5 Gm
Ipecac. Pulv., ...	5 Gm
Camphora, ...	5 Gm

Make 100 pills.

No. 191.

℞. Ipecac. Pulv., ...	1 Gm
Opium Pulv., ...	2 Gm
Pil. Hydrarg., ...	2 Gm

Make 100 pills.

No. 192.

R. PLUMB. ACET., ... 3 Gm
MORPH. ACET., ... 1.50 Gm
EXTR. HYOSCYAM., ... 15 Gm
Make 100 pills.

No. 193.

R. OPIUM PULV., ... 2 Gm
ACID. TANNIC., ... 15 Gm
Make 100 pills.

No. 194.

R. PLUMB. ACET., ... 10 Gm
OPIUM PULV., ... 5 Gm
Make 100 pills.

No. 195.

R. ARGENT. NITR., ... 2 Gm
OPIUM PULV., ... 2 Gm
Make 100 pills.

No. 196.

R. OPIUM PULV., ... 1 Gm
CUPR. SULPH., ... 1.50 Gm
EXTR. GENT., ... 10 Gm
Make 100 pills.

No. 197.

R. OPIUM PULV., ... 1.50 Gm
DIGITAL. PULV., ... 4 Gm
QUIN. SULPH., ... 10 Gm
Make 100 pills.

No. 198.

R. EXTR. COLCH. ACET., ... 5 Gm
PULV. IPECAC. COMP.,. . . . . 20 Gm
Make 100 pills.

No. 199.

R. EXTR. COLCH. ACET., ... 5 Gm
DIGITALIS PULV., ... 3 Gm
EXTR. COLOC. COMP., ... 1 Gm
Make 100 pills.

No. 200.

R. EXTR. COLCH. ACET., ... 10 Gm
MASS. PIL. HYDRARG., ... 15 Gm
Make 100 pills.

No. 201.

R. ALOE PURIFIC., ... 6.50 Gm
IPECACUANHA PULV., ... 6.50 Gm
Make 100 pills.

No. 202.

R. EXTR. COLOC. COMP., ... 20 Gm
EXTR. HYOSCYAM., ... 10 Gm
Make 100 pills.

No. 203.

R. EXTR. BELLAD., ... 2 Gm
EXTR. GENT., ... 5 Gm
EXTR. COLOC. COMP., ... 10 Gm
ÆTHEROL. CARUM., ... 1 Gm
Make 100 pills.

No. 204.

R. Resina Podoph., ... 1.50 Gm
   Extr. Hyosycam., ... 7.50 Gm
   Pulv. Capsicum, ... 7.50 Gm
Make 100 pills.

No. 205.

R. Resina Podoph., ... 1.50 Gm
   Extr. Nux Vom., ... 2 Gm
   Extr. Bellad., ... 2 Gm
Make 100 pills.

No. 206.

R. Resina Podoph., ... 2 Gm
   Extr. Nux Vom., ... 1 Gm
   Extr. Bellad., ... 1 Gm
   Extr. Coloc. Comp., ... 20 Gm
Make 100 pills.

No. 207.

R. Resina Podoph., ... 3 Gm
   Extr. Nux Vom., ... 1.50 Gm
   Extr. Bellad., ... 1.50 Gm
   Aloe Socotr. Pulv., ... 5 Gm
Make 100 pills.

No. 208.

R. Extr. Bellad., ... 1.50 Gm
   Extr. Nux Vom., ... 3 Gm
   Ipecac. Pulv., ... 5 Gm
   Extr. Coloc. Comp., ... 15 Gm
Make 100 pills.

No. 209.

℞. PIL. HYDRARG., ... 6.50 Gm
    RESINA SCAMM., ... 6.50 Gm
    ALOE SOCOTR. PULV., ... 6.50 Gm
    ÆTHEROL. CARUM, ... 1 Gm
Make 100 pills.

No. 210.

℞. CAMBOG. PULV., ... 1.50 Gm
    HYDRARG. CHLOR. MIT., ... 6 Gm
    EXTR. JALAPA, ... 6 Gm
    EXTR. COLOC. COMP.; . . . . 8 Gm
Make 100 pills.

No. 211.

℞. EXTR. COLOC. COMP., ... 10 Gm
    HYDR. CHLOR. MIT., ... 10 Gm
    EXTR. HYOSCYAM., ... 2.50 Gm
    EXTR. NUX VOM., ... 1 Gm
    ALOE PULV., ... 1 Gm
    IPECAC. PULV., ... 1 Gm
Make 100 pills.

No. 212.

℞. EXTR. ALOE, ... 2.50 Gm
    QUINID. SULPH., ... 12.50 Gm
    FERR. SULPH. EXSICC., . . . 6 Gm
    EXTR. BELLAD., ... 1.20 Gm
    EXTR. NUX VOM., ... 1.30 Gm
Make 100 pills.

No. 213.

℞. FERR. SULPH., ... 10 Gm
    POTASS. CARB. PUR., . . . 10 Gm
Make 100 pills.

No. 214.

R. CINCHONID. SULPH., ... 5 Gm
QUINID. SULPH., ... 5 Gm
RESINA LEPTANDRA, ... 10 Gm
RESINA PODOPH., ... 2 Gm
IPECAC. PULV., ... 1 Gm

Make 100 pills.

No. 215.

R. CINCHONID. SULPH., ... 8 Gm
POTASS. IODIDUM, ... 8 Gm
ACID CITRICUM, ... 8 Gm

Make 100 pills.
(Surgeon John Vansant, U. S. M. H. S.)

No. 216.

R. FERR. REDUCTUM, ... 7 Gm
QUININ. SULPH., ... 8 Gm
STRYCHNIN. NITR., ... 0.15 Gm

Make 100 pills.

No. 217.

R. FERR. REDUCT., ... 7 Gm
QUINID. SULPH., . . . 8 Gm
ACID. ARSENIOS., ... 0.15 Gm

Make 100 pills.

No. 218.

℞. Acid. Arsenios., ...	0.20 Gm
Strychnin. Nitr., ...	0.20 Gm
Quinidin. Sulph., ...	5 Gm
Ferrum Reductum, ...	7 Gm

Make 100 pills.

No. 219.

℞. Quin. Sulph., ...	15 Gm
Mass. Pil. Hydrarg., . . .	8 Gm

Make 100 pills.

CONFECTIONS.

No. 220.

℞. Sulphur. Lotum, ...	15 Gm
Potass. Bitartr., ...	15 Gm
Confectio Sennæ, ...	60 Gm
Syrupus Sacch., ...	10 Gm
	100 Gm

Mix.

No. 221.

℞. Rheum Pulv., ...	2 Gm
Myrist. Pulv., ...	1 Gm
Guaiac. Resinæ Pulv., . . .	1 Gm
Sulph. Lotum, ...	16 Gm
Potass. Bitartr., ...	32 Gm
Mel. Despum., ...	48 Gm
	100 Gm

Mix.

HYPODERMIC INJECTIONS.

No. 222.

R. MORPH. ACET., ... 15 Gm
AQUA, sufficient
to make . . . . . . 100 fGm
Make a solution.

No. 223.

R. ATROPIN. SULPH., ... 0.35 Gm
AQUA, sufficient
to make . . . . . . 100 fGm
Make a solution.
Dose, 1½ to 3 fluidimes.

No. 224.

R. IODUM, ... 8 Gm
ALCOHOL ABSOL., sufficient
to make . . . . . . 100 fGm
Make a solution. Dose, 0.50 to 1 fGm.

ENEMATA.

No. 225.

R. SODIC. CHLORID., ... 30 Gm
DECOCT. HORD., ... 600 fGm
Mix.

No. 226.

R. OLEUM OLIVA, ... 100 fGm
DECOCT. HORD., ... 500 fGm
Mix.

No. 227.

R. Oleum Ricinus, ... 30 fGm
Decoct. Hord., ... 500 fGm
Mix.

No. 228.

R. Tinct. Asafœtida, ... 15 fGm
Mucil. Amylum, . . . ! 600 fGm
Mix.

No. 229.

R. Ætherol. Terebinth., . . . 15 fGm
Ol. Oliva, ... 50 fGm
Sapo, ... 25 Gm
Aqua Fervida, ... 500 fGm
Sodic. Chlorid., ... 15 Gm
Mix.

No. 230.

R. Quinin. Hydrobrom., . . . 0.60 Gm
Alcohol, ... 0.50 fGm
Mucil. Amylum, ... 10 fGm
Aqua, ... 5 fGm
Mix.

No. 231.

R. Tinct. Opium, ... 1 fGm
Mucil. Amylum, ... 60 fGm
Mix.

No. 232.

R. Bismuth. Subcarb., ... 50 Gm
Extr. Opium, ... 0.20 Gm
Glycerinum, ... 100 fGm
Aqua, ... 100 fGm
Mix. 40 fGm to be used at a time.

No. 233.

R. STRONG BEEF TEA, ... 90 fGm
   BRANDY, ... 30 fGm
Mix.

SUPPOSITORIES.*
No. 234.

R. EXTR. CONIUM, ... 0.60 Gm
   OL. THEOBROMA, ... 2 fGm
Mix.

No. 235.

R. EXTR. BELLAD., ... 0.12 Gm
   OL. THEOBROMA, ... 1 fGm
Mix.

No. 236.

R. MORPH. HYDROCHLOR., ... 0.03 Gm
   OL. THEOBROMA, ... 2 fGm
Mix.

·   No. 237.

R. ATROPIN. SULPH., ... 0.003 Gm
   OL. THEOBROMA, ... 3 fGm
Mix.

No. 238.

R. ATROPIN. SULPH., ... 0.003 Gm
   MORPH. ACET., ... 0.03 Gm
   OL. THEOBROMA, ... 3 fGm
Mix.

---

* The weight of suppositories should depend somewhat upon the
active constituent, and may, of course, be fixed in each case according
to circumstances. For the officinal suppositories of the U. S. P. a uni-
form weight of 30 grains (2 grams) is prescribed.

No. 239.

R. POTASS. BROMID., ... 0.60 Gm
EXTR. BELLAD., ... 0.06 Gm
OL. THEOBROMA, ... 2 fGm
Mix.

No. 240.

R. ACID. TANN., ... 0.20 Gm
EXTR. BELLAD., ... 0.12 Gm
OL. THEOBROMA, ... 2 fGm
Mix.

No. 241.

R. ACID. TANNIC., ... 0.60 Gm
OL. THEOBROMA, ... 3 fGm
Mix.

No. 242.

R. ACID. CARBOLIC., ... 0.06 Gm
OL. THEOBROMA, ... 2 fGm
Mix.

No. 243.

R. IODOFORM., ... 0.10 Gm
OL. THEOBROMA, ... 1 fGm
Mix.

C.

EXTERNAL REMEDIES.

COLLYRIA.

No. 244.

R. ALUMEN, ... 1 Gm
AQUA, ... 100 fGm
Make a solution, and filter.

No. 245.

R. ARGENT. NITR., ... 0.20 Gm
AQUA DESTILL., ... 100 fGm
Make a solution, and filter.

No. 246.

R. ATROPIN. SULPH., ... 0.04 Gm
AQUA DESTILL., ... 100 fGm
Make a solution, and filter.

No. 247.

R. SODIC. BIBOR., ... 2 Gm
AQUA ROSA, ... 100 fGm
Make a solution, and filter.

No. 248.

R. PLUMB. ACET., ... 0.40 Gm
AQUA ROSA, ... 100 fGm
Make a solution, and filter.

No 249.

R. ZINCIC. SULPH., ... 0.40 Gm
AQUA, ... 100 fGm
Make a solution, and filter.

NASAL DOUCHES.

Three hundred fluigrams is usually sufficient. The temperature of the liquid should be about 32° C. (90° F.).

No. 250.

R. ALUMEN., ... 1 Gm
AQUA, ... 100 fGm
Make a solution, and filter.

No. 251.

R. ACID. TANN., ... 0.60 Gm
AQUA, ... 100 fGm
Make a solution, and filter.

No. 252.

R. POTASS. PERMANGAN., ... 0.01 Gm
AQUA, ... 100 fGm
Make a solution, and filter.

No. 253.

R. ACID. BORIC., ... 0.60 Gm
AQUA, ... 100 fGm
Make a solution, and filter.

No. 254.

R. ZINCIC. SULPHO-CARBOL., . . . 0.40 Gm
AQUA, ... 100 fGm
Make a solution, and filter.

No. 255.

R. QUIN. SULPH., ... 0.10 Gm
AQUA, ... 100 fGm
Make a solution. This has been used in hay-fever, a small quantity being snuffed up into the nose from the palm of the hand.

VAPORS AND INHALATIONS.

No. 256.

R. ÆTHER. FORT., ... 50 fGm
ALCOHOL FORT., ... 50 fGm
Mix. Use 5 fGm in 500 fGm water for each inhalation.

No. 257.

R. ACID. ACET., ... 50 fGm
   ACID. ACET. GLAC., ... 50 fGm
Mix.   Use 10 fGm in 500 fGm water at 60° C. (140° F.)
for each inhalation.

No. 258.

R. THYMOL, ... 1.00 Gm
   ALCOHOL, ... 10 fGm
   MAGNES. CARB., ... 0.50 Gm
   AQUA, sufficient
      to make, ... 100 fGm
Mix.   Use 5 fGm in 500 fGm water at 60 C.° (140° F.)
for each inhalation.

No. 259.

R. CREASOTUM, ... 2 fGm
   AQUA FERVIDA, ... 500 fGm
Mix.

No. 260.

R. ACID. CARBOL. CRYST. PUR., . . 87.50 Gm
   AQUA, ... 12.50 fGm
Mix.   Use 20 drops in 500 fGm Water at 60° C. (140°
F.) ; or 8 fGm at 25° C. to 40° C. (80° to 100° F.) or 5 fGm
for dry inhalation.

No. 261.

CARBOLIC ACID SPRAY.

R. ACID. CARBOL. CRYST., . . . 4 Gm
   AQUA, ... 600 fGm
Mix.

No. 262.

℞. Liqu. Iodum. Comp., ... 1 fGm
    Infus. Humulus, ... 120 fGm
Mix.

No. 263.

℞. Tinct. Iodum, ... 5 fGm
    Aqua, ... 40 fGm
Mix.

No. 264.

℞. Calx Chlorata, ... 50 Gm
Aqua, sufficient to moisten it well.

No. 265.

℞. Extr. Conium, ... 5 fGm
    Liqu. Potassa, ... 5 fGm
    Aqua, ... 50 fGm
Mix.

No. 266.

℞. Oleum Cubeba, ... 8 fGm
    Magnes. Carb., ... 4 Gm
    Aqua, ... 100 fGm
Mix.

No. 267.

℞. Acid. Salicyl., ... 15 Gm
    Sod. Bibor., ... 15 Gm
    Aqua Fervida, ... 500 fGm
Make a solution.

GARGLES.

No. 268.

℞. Potass. Permangan.,	.	.	.		1 Gm
Aqua, sufficient					
to make,	.	.	.	.	100 fGm

Make a solution.

No. 269.

℞. Potass. Chlor.,	.	.	.	.	5 Gm
Aqua, sufficient					
to make,	.	.	.	.	100 fGm

Make a solution.

No. 270.

℞. Alumen.,	.	.	.	.	1 Gm
Potass. Chlor.,	.	.	.	.	1 Gm
Aqua, sufficient					
to make,	.	.	.	.	100 fGm

Make a solution.

No. 271.

℞. Acid. Tann., .	.	.	.	.	3 Gm
Potass. Chlor.,	.	.	.	.	1.50 Gm
Glycerinum, .	.	.	.	.	25 fGm
Aqua, sufficient					
to make,	.	.	.	.	100 fGm

Mix.

No. 272.

℞. Alumen.,	.	.	.	.	1 Gm
Acid. Tannic., .	.	.	.	.	1.50 Gm
Aqua, sufficient					
to make,	.	.	.	.	100 fGm

Mix.

No. 273.

R. Sod. Bibor.,	. . . . .	5 Gm
Glycerinum,	. . . . .	5 fGm
Tinct. Myrrha,	. . . .	5 fGm
Aqua,		90 fGm
Mix.		100 fGm

No. 274.

R. Acid. Tannic.,	. . . .	3 Gm
Glycerinum,	. . . . .	10 fGm
Mist. Camphora,	. . . .	90 fGm
Mix.		100 fGm

No. 275.

R. Acid. Carbol. Cryst.,	. . .	0.40 Gm
Glycerinum,	. . . . .	10 fGm
Aqua,	. . . . . .	90 fGm
Mix.		100 fGm

No. 276.

R. Acid. Acetic.,	. . . . .	3 fGm
Glycerinum,	. . . . .	5 fGm
Aqua,	. . . . . .	92 fGm
Mix.		100 fGm

No. 277.

R. Acid. Hydrochlor. Dil.,	. .	3 fGm
Glycerinum,	. . . . .	7 fGm
Aqua,	. . . . . .	90 fGm
Mix.		100 fGm

No. 278.

R. Acid. Nitr. Dil.,	. . . .	1 fGm
Glycerinum,	. . . . .	9 fGm
Aqua,	. . . . . .	90 fGm
Mix.		100 fGm

No. 279.

R. Liq. Soda Chlorata,   .   .   .   6 fGm
  Aqua, ...   94 fGm
Mix.   100 fGm

No. 280.

R. Tinct. Ferr. Chlor.,   .   .   .   6 fGm
  Potass. Chlor. Pulv.,   .,   .   .   5 Gm
  Aqua, sufficient
    to make, ...   100 fGm
Make a solution.

INJECTIONS.

No. 281.

R. Acid. Tannic., ...   3 Gm
  Glycerinum, ...   25 fGm
  Aqua, sufficient
    to make, ...   100 fGm
Mix.

No. 282.

R. Zincic. Sulph., ...   0.40 Gm
  Morph. Sulph., ...   0.10 Gm
  Aqua, sufficient
    to make, ...   100 fGm
Make a solution.

No. 283.

R. Zincic. Sulph., ...   0.40 Gm
  Morphin. Sulph., ...   0.10 Gm
  Bol. Armen. Pulv. Subtiliss.,   .   .   0.80 Gm
  Aqua, sufficient
    to make, ...   100 fGm
Mix.

No. 284.

℞. Liqu. Soda Chlorata,   .   .   .   1 fGm
Aqua, .   •   •   •   •   •   .   99 fGm
                                     100 fGm
Mix.

No. 285.

℞. Quinin. Sulph., ...   0.15 Gm
Mucil. Acacia,   .   .   .   •   .   50 fGm
Aqua, ...   50 fGm
                                     100 fGm
Mix.

No. 286.

℞. Strychn. Nitr., ...   0.06 Gm
Acid. Acet. Dil., ...   0.20 fGm
Aqua, sufficient
    to make, ...   100 fGm
Make a solution.

LOTIONS AND PAINTS.

No. 287.

℞. Ammonic. Chlorid., ...   5 Gm
Acid. Acetic., ...   10 fGm
Alcohol, ...   10 fGm
Aqua, sufficient
    to make, ...   100 fGm
Mix.

No. 288.

℞. Liqu. Soda Chlorata,   .   .   .   15 fGm
Aqua,   .   .   .   ,   ,   .   85 fGm
                                     100 fGm
Mix.

No. 289.

R. SOL. ZINCIC. CHLORID., . . . 1 fGm
AQUA, ... 99 fGm
————
100 fGm
Mix.

No. 290.

R. BROMUM, ... 0.75 fGm
POTASSIC. BROMID., ... 1.50 Gm
AQUA, sufficient
to make, ... 100 fGm
Make a solution.

No. 291.

R. BROMUM, ... 20 fGm
POTASS. BROMID., ... 6 Gm
AQUA, sufficient
to make, ... 100 fGm
Make a solution.

No. 292.

R. IODUM, ... 25 Gm
POTASS. IODID., ... 50 Gm
AQUA, sufficient
to make, ... 100 fGm
Make a solution.

No. 293.

R. IODUM, ... 3 Gm
BROMUM, ... 6 fGm
ALCOHOL FORT., sufficient
to make, ... 100 fGm
Make a solution.

No. 294.

R. Iodum, ... 10 Gm
Iodoformum, ... 10 Gm
Æther. Fort., sufficient
    to make, ... 100 fGm
Make a solution.

No. 295.

R. Iodoformum, ... 20 Gm
Æther. Fort., sufficient
    to make, ... 100 fGm
Mix.

No. 296.

R. Sodic. Hyposulph., ... 20 Gm
Acid. Carbol., ... 1 Gm
Glycerinum, ... 50 fGm
Aqua Rosa, sufficient
    to make, ... 100 fGm
Mix.

No. 297.

R. Collodium, ... 90 fGm
Oleum Ricinus, ... 5 fGm
Acid. Carbol. Cryst., . . . 5 Gm
Mix.   100 fGm

No. 298.

R. Acid. Carbol. Cryst., . . . 2 Gm
Ol. Linum, sufficient
    to make . . . . . 100 fGm
Mix.

No. 299.

R. Acid. Carbol., ... 10 Gm
Oleum Linum, sufficient
    to make . . . . . 100 fGm
Mix.

No. 300.

R. OLEUM TIGLIUM, ...   15 fGm
ÆTHER., ...   30 fGm
TINCT. IODUM COMP.,   .   .   .   55 fGm
Mix.   100 fGm

No. 301.

R. OLEUM TIGLIUM, ...   50 fGm
COLLOD. FLEXILE, ...   50 fGm
Mix.   100 fGm

No. 302.

R. PLUMBIC. SUBACET., ...   2 Gm
TINCT. OPIUM, ...   3 fGm
AQUA, sufficient
   to make, ...   100 fGm
Mix.

No. 303.

R. ACID. TANNIC, ...   50 Gm
GLYCERINUM, sufficient
   to make, ...   100 fGm
Make a solution.
Each fGm represents 0.50 Gm of tannin.

No. 304.

R. ACID. TANN., ...   2 Gm
TINCT. LAVEND. COMP.,   .   .   .   15 fGm
TINCT. OPIUM, ...   8 fGm
AQUA, sufficient
   to make, ...   100 fGm
Mix.

No. 305.

R. Tinct. Iodum Comp.,   .   .   .   40 fGm
   Chloroformum, ...   20 fGm
   Tinct. Aconitum Rad.,   .   .   .   40 fGm
Mix.   100 fGm

LINIMENTS.

No. 306.

R. Linim. Sapo, ...   80 fGm
   Ætherol. Terebinth.,   .   .   .   20 fGm
Mix.   100 fGm

No. 307.

R. Chloroform., ...   20 fGm
   Tinct. Opium, ...   10 fGm
   Linim. Sapo, ...   70 fGm
Mix.   100 fGm

No. 308.

R. Chloroform., ...   30 fGm
   Tinct. Aconit. Rad.,   .   .   .   5 fGm
   Aqua Ammonia, ...   5 fGm
   Oleum Oliva, ...   60 fGm
Mix.   100 fGm

No. 309.

R. Chloral. Hydr., ...   20 Gm
   Camphora, ...   20 Gm
   Tinct. Aconit. Rad.,   .   .   .   20 fGm
   Tinct. Opium, ...   20 fGm
   Alcohol, ...   10 fGm
   Ætherol. Mentha Pip.,   .   .   .   10 fGm
Mix.   100 fGm

No. 310.

R. Liqu. Calx, ... 30 fGm
   Oleum Linum, ... 70 fGm
Mix.   100 fGm

OINTMENTS.

No. 311.

R. Acid. Boric., ... 15 Gm
   Butyrum Petroleum, . . . 85 Gm
Mix.   100 Gm

No. 312.

R. Acid. Carbol. Cryst., . . . 5 Gm
   Butyrum Petroleum, . . . 95 Gm
Mix.   100 Gm

No. 313.

R. Bals. Peruv., ... 10 Gm
   Glycerinum, ... 5 Gm
   Ungu. Resina, ... 87 Gm
Mix.   100 Gm

No. 314.

R. Potassic. Iodid., ... 20 Gm
   Ungu. Stramon., ... 80 Gm
Mix.   100 Gm

No. 315.

R. Extr. Hyoscyam., ... 10 Gm
   Acid. Tannic., ... 10 Gm
   Butyrum Petroleum, . . . 80 Gm
Mix.   100 Gm

No. 316.

R. Ungu. Hydrarg. Nitr., . . . 10 Gm
    Acid. Tannic., ... 10 Gm
    Extr. Hyoscyam., . . . 10 Gm
    Butyrum Petroleum, . . 70 Gm
Mix.   100 Gm

No. 317.

R. Sulphur. Præcip., ... 15 Gm
    Pyroleum Cadinum, ... 15 Gm
    Creta Præparata, ... 10 Gm
    Sapo Mollis Virid., ... 30 Gm
    Butyrum Petroleum, . . 30 Gm
Mix.   100 Gm

No. 318.

R. Sulph. Præcip., ... 20 Gm
    Potass. Carb., ... 10 Gm
    Butyrum Petroleum, . . 70 Gm
Mix.   100 Gm

No. 319.

R. Pix Liquida, ... 15 Gm
    Sulphur Præcip., ... 15 Gm
    Creta Præpar., . . . 10 Gm
    Sapo Moll. Virid., ... 30 Gm
    Butyr. Petrol., . . . 30 Gm
Mix.   100 Gm

No. 320.

R. Zincic. Sulpho-carbol., . . . 10 Gm
    Butyr. Petrol., . . . 90 Gm
Mix.   100 Gm

No. 321.

R. Ungu. Hydrarg. Nitr., . . . 20 Gm
   Butyr. Petrol., ... 80 Gm
Mix.   100 Gm

No. 322.

R. Iodoform. Pulv., ... 2 Gm
   Butyr. Petrol., ... 98 Gm
Mix.   100 Gm

No. 323.

R. Iodoform. Pulv., ... 20 Gm
   Butyr. Petrol., ... 80 Gm
Mix.   100 Gm

No. 324.

R. Atropin. Sulph., ... 0.50 Gm
   Extr. Opium, ... 1 Gm
   Iodoformum, ... 4.50 Gm
   Butyr. Petrol., . . . 95 Gm
Mix.   100 Gm

PLASTERS.

No. 325.

R. Extr. Bellad., ... 30 Gm
   Empl. Hydrarg., ... 35 Gm
   Empl. Plumb., ... 35 Gm
Mix.

POWDERS FOR EXTERNAL USE.

No. 326.

R. Cupric. Sulph. Pulv., . . . 50 Gm
   Cinchona Cort. Pulv., . . . 50 Gm
Mix.   100 Gm

BATHS.

No. 327.

R. ACID. NITRIC., ... 400 fGm
ACID. HYDROCHLOR., ... 800 fGm
AQUA, ... 120 Liters
Mix.

No. 328.

R. ACID. NITRIC., ... 50 fGm
ACID. HYDROCHLOR;, ... 35 fGm
AQUA, ... 120 Liters
Mix.

No. 329.

R. SODIC. CARB. CRYST., ... 200 Gm
AQUA, ... 120 Liters
Mix.

No. 330.

R. SAL. MARINUM, ... 4 Kilos
AQUA, ... 120 Liters
Mix.

No. 331.

R. POTASSA SULPHUR., ... 200 Gm
AQUA, ... 120 Liters
Mix.

MERCURIAL FUMIGATION (Mercurial Vapor Bath).

No. 332.

R. HYDRARG. OXID. NIGR., . . . 5 Gm
HYDRARG. OXID. RUBR., . . . 5 Gm
HYDRARG. SULPHURET. RUBR., . . 5 Gm
Mix.

CHLORINE FUMIGATION.

No. 333.

℞. BLACK OXIDE OF MANAGANESE IN POWDER,

fifty grams, ... 50

HYDROCHLORIC ACID, CRUDE, one hundred

and fifty fluigrams, ... 150

Mix on a soup plate or other suitable vessel.

COPPERAS AND CARBOLIC ACID DISINFECTANT.

No. 334.

℞. FERR. SULPH. CRUD., ... 2 Kilos

ACID. CARBOL. CRUD., . . . 300 fGm

AQUA, ... 15 Liters

Make a solution.

LISTER'S ANTISEPTIC TREATMENT.

a. Before and during Operation.

1. Carbolic Acid 'Spray.—Steam passing through a solution of 1 part of carbolic acid in 30 parts of water. As it issues from the jet, the solution contains about 1 part of acid in 40 of water.

2. Sponges, hands of operators, etc., dipped in solution of carbolic acid : 1 in 20.

3. Instruments covered with oil containing one-tenth part carbolic acid : some are dipped into or kept in watery solution : 1 in 20.

4. During intermission of spray, the wound is covered with a cloth dipped in carbolic acid solution : 1 in 20.

b. After Operation.

1. A strip of lint soaked in an oily solution of carbolic acid (1 in 10), or a pure rubber drainage-tube, similarly treated, is left hanging from the wound during the first (and if necessary following) days. Either of them are cut off flush with the edge of the wound.

2. Over this is placed the Protective, into which a small hole is cut, corresponding with the end of the drainage-tube. The Protective consists of a layer of oiled silk, coated on both sides with copal varnish, and afterwards brushed over with dextrin, which latter enables it to become uniformly moistened when dipped into a solution of carbolic acid (1 in 40). It is thus immersed just before being laid upon the wound, and is intended to prevent irritation which would be caused by the actual contact of the antiseptic dressing with the wound.

3. Two or three layers of gauze dipped in a watery solution of carbolic acid (1 in 40) are next applied. Then,

4. Seven layers of the Antiseptic Gauze, being a cotton fabric of open texture, impregnated with a mixture of 5 parts resin, 7 parts paraffin, and 1 part carbolic acid.

5. Over this is applied the mackintosh, which is about 1 inch less in size than the gauze.

6. Then another layer of antiseptic gauze is applied, and finally,

7. Carbolized bandages, sufficient to retain the dressings, etc.

[CHARLES RICE.]

PART III.

POSOLOGICAL TABLE.

DOSE TABLE.

Remedies.	Dose in Apothecaries' Weights and Measures.	Dose in Metric Terms.
Absinthium	15 to 60 gr.	1 to 4 Gm.
Absinth. Tinct.	30 to 60 min.	2 to 4 fGm.
Acetum	1 to 2 fl. dr.	5 to 10 fGm.
Acet. Lobel.	15 to 60 min.	1 to 4 fGm.
Acet. Opium	5 to 10 min.	3 to 6 f D.
Acet. Sanguin. Alter.	15 to 30 min.	1 to 2 fGm.
" " Emet.	1 to 4 fl. dr.	5 to 15 fGm.
Acet. Scilla	8 to 30 min.	5 to 20 f D.
Achillea	30 to 60 gr.	2 to 4 Gm.
Acid, Acet.	5 to 15 min.	3 to 10 f D.
" " Dil	1 to 2 fl. dr.	5 to 10 fGm.
" Arsenios.	$\frac{1}{64}$ to $\frac{1}{12}$ gr.	1 to 5 Mills.
" Benzoic.	8 to 15 gr.	50 cts. to 1 Gm.
" Boric. .	15 to 60 gr.	1 to 4 Gm. .
" Carbolic	$\frac{1}{2}$ to 3 gr.	3 to 20 cents.
" Chrysophan.	5 to 15 gr.	30 cts. to 1 Gm.
" Citric	5 to 30 gr.	30 cts. to 2 Gm.
" Gallic	3 to 15 gr.	20 cts. to 1 Gm.
" " in albuminuria	10 to 60 gr.	60 cts. to 4 Gm.
" Gallo-Tannic.	2 to 10 gr.	10 to 60 cents.
" Hydriodic. Dil.	15 to 30 min.	1 to 2 fGm.
" Hydrobrom. 34 p. c.	10 to 15 gr.	60 cts. to 1 Gm.
" " Dil. 10 p.c.	20 to 40 min.	1.50 to 2.50 fGm.
" Hydrochlor.	5 to 15 min.	3 to 10 f D.
" " Dil.	20 to 60 min.	15 to 40 f D.
" Hydrocyan. Dil.	2 to 6 min.	1 to 4 f D.
" Lactic. sp. gr. 1.22	1 to 2 fl. dr.	5 to 10 fGm.
" Muriat. Dil.	10 to 30 min.	5 to 20 f D.
" Nitric	5 to 8 min.	3 to 5 f D.
" " Dil.	20 to 40 min.	15 to 30 f D.
" Nitromur.	2 to 5 min.	1 to 3 f D.

Remedies.	Dose in Apothecaries' Weights and Measures.	Dose in Metric Terms.
Acid. Nitromur. Dil.	10 to 20 min.	5 to 15 f D.
" Oxalic	$\frac{1}{2}$ to 1 gr.	3 to 5 cents.
" Phosph. Glac.	1 to 2 gr.	5 to 10 cents.
" " Dil.	20 to 60 min.	15 to 40 f D.
" Picric	$\frac{1}{50}$ to $\frac{1}{10}$ gr.	1 to 5 Mills.
" Salicyl.	8 to 40 gr.	0.50 to 2.50 Gm.
" Sulphuric	1 to 5 min.	1 to 3 f D.
" " Dil.	8 to 40 min.	5 to 25 f D.
" " Arom.	5 to 30 min.	3 to 20 f D.
" Sulphuros.	20 to 60 min.	15 to 40 f D.
" Tannic	2 to 10 gr.	10 to 60 cents.
" Tartaric	30 to 60 gr.	2 to 5 Gm.
Aconitin (eclectic; see next line)	$\frac{1}{24}$ to $\frac{1}{12}$ gr.	2 to 6 mills.
Aconitina (alkaloid)	$\frac{1}{200}$ to $\frac{1}{100}$ gr.	0.3 to 0.6 mills.
Aconit. Fol.	1 to 3 gr.	5 to 20 cents.
" " Extr.	$\frac{1}{6}$ to $\frac{1}{3}$ gr.	1 to 2 cents.
" " Tinct.	5 to 10 min.	3 to 6 f D.
" Rad.	$\frac{1}{2}$ to 2 gr.	3 to 15 cents.
" " Extr.	$\frac{1}{10}$ to $\frac{1}{6}$ gr.	5 to 10 mills.
" " Tinct.	3 to 5 min.	1 to 3 f D.
" " " Fleming's	1 to 3 min.	0.50 to 2 f D.
Aesculus Glabra	2 to 8 dr.	10 to 30 Gm.
" Hippocast.	15 to 50 gr.	1 to 3 Gm.
Aether Acet.	10 to 60 min.	5 to 40 f D.
" Fort.	30 to 60 min.	2 to 4 f Gm.
" Nitros. Spir.	$\frac{1}{2}$ to 2 fl. dr.	2 to 10 f Gm.
" Spirit. Comp.	$\frac{1}{2}$ to 2 fl. dr.	2 to 10 f Gm.
Aetheroleum Am. Am.	$\frac{1}{10}$ to $\frac{1}{2}$ min.	0.05 to 0.3 f D.
" Anisum	5 to 15 min.	3 to 10 f D.
" Anthemis	3 to 10 min.	2 to 6 f D.
" Aurant.	1 to 5 min.	1 to 3 f D.
" Cajuput	1 to 5 min.	1 to 3 f D.
" Camphora	1 to 3 min.	1 to 2 f D.
" Carum	1 to 10 min.	1 to 6 f D.
" Caryoph.	1 to 5 min.	1 to 3 f D.
" Cassia	1 to 3 min.	1 to 2 f D.

Remedies.	Dose in Apothecaries' Weights and Measures.	Dose in Metric Terms.
Aetheroleum Chenopod.	4 to 8 min.	2 to 5 f D.
" Cinnam.	1 to 3 min.	1 to 2 f D.
" Copaiba	10 to 15 min.	5 to 10 f D.
" Coriandrum	2 to 10 min.	1 to 6 f D.
" Cubeba	10 to 20 min.	5 to 10 f D.
" Eriger. Can.	5 to 10 min.	3 to 6 f D.
" Eucalypt.	10 to 20 min.	5 to 10 f D.
" Fœnicul.	5 to 15 min.	3 to 10 f D.
" Gaulther.	2 to 10 min.	1 to 6 f D.
" Hedeoma	2 to 10 min.	1 to 6 f D.
" Juniperus	5 to 10 min.	3 to 6 f D.
" Lavendula	1 to 5 min.	1 to 3 f D.
" Limon.	1 to 5 min.	1 to 3 f D.
" Menth. Pip.	1 to 3 min.	1 to 2 f D.
" " Vir.	1 to 5 min.	1 to 3 f D.
" Monarda	1 to 5 min.	1 to 3 f D.
" Myristica	1 to 3 min.	1 to 2 f D.
" Petroleum Rectif.	10 to 60 min.	5 to 40 f D.
" Pimenta	1 to 5 min.	1 to 3 f D.
" Rosmar.	1 to 5 min.	1 to 3 f D.
" Ruta	1 to 5 min.	1 to 3 f D.
" Sabina	1 to 3 min.	1 to 2 f D.
" Santal.	15 to 30 min.	1 to 2 f Gm.
" Sassafras	2 to 10 min.	1 to 6 f D.
" Succin. Rect.	5 to 15 min.	3 to 10 f D.
" Terebinth.	5 to 30 min.	3 to 20 f D.
" " as a tænicide	½ to 2 fl. dr.	2 to 8 f Gm.
" Valeriana	1 to 5 min.	1 to 3 f D.
Agrimonia	½ to 2 dr.	2 to 8 Gm.
Ailanthus	8 to 30 gr.	0.50 to 2 Gm.
Alcohol Amyl.	2 to 8 min.	1 to 5 f D.
Aletrin (eclectic)	½ to 2 gr.	3 to 10 cents.
Aletris	8 to 30 gr.	0.50 to 2 Gm.
Allium	½ to 2 dr.	2 to 8 Gm.
" Syrupus	½ to 1½ fl. dr.	2 to 6 f Gm.
Alnuin (eclectic)	2 to 10 gr.	10 to 60 cents.

Remedies.	Dose in Apothecaries' Weights and Measures.	Dose in Metric Terms.
Alnus	10 to 60 gr.	50 cts. to 40 Gm.
Aloinum	⅙ to 3 gr.	1 to 20 cents.
Aloe Barb. lax.	2 to 3 gr.	10 to 20 cents.
" " purg.	10 to 20 gr.	50 to 120 cents.
" Cap.	Same as Aloe Barb.	————
" Soc.	Same as Aloe Barb.	————
" Decoct. Co.	4 to 16 fl. dr.	15 to 60 f Gm.
" Extr. lax.	½ to 8 gr.	3 to 20 cents.
" " purg.	3 to 10 gr.	15 to 60 cents.
" Purificata	2 to 4 gr.	10 to 25 cents.
" Pil.	1 to 3 pills	1 to 3 pills.
" " Asafœt.	2 to 5 pills	1 to 3 pills.
" " Ferr.	5 to 10 gr.	30 to 60 cents.
" " Mast.	1 to 2 pills	1 to 2 pills.
" " Myrrha	3 to 6 pills	3 to 6 pills.
" Tinct. lax.	1 to 2 fl. dr.	4 to 8 f Gm.
" " purg.	4 to 12 fl. dr.	15 to 50 f Gm.
" Tinct. Myrrha	1 to 2 fl. dr.	4 to 8 f Gm.
" Vinum stomach.	1 to 2 fl. dr.	4 to 8 f Gm.
" " purg.	4 to 8 fl. dr.	15 to 30 f Gm.
Alstonia	1 to 4 dr.	4 to 8 Gm.
Althæa	½ to 2 dr.	2 to 8 Gm.
" Decoct.	*Ad libitum*	*Ad libitum*
" Syrup.	½ to 2 fl. dr.	2 to 10 f Gm.
Alumen (pot. or ammon.)	5 to 20 gr.	30 to 120 cents.
" Exsicc.	5 to 10 gr.	30 to 60 cents.
Ammoniacum	5 to 10 gr.	30 to 60 cents.
" Mixt.	4 to 8 fl. dr.	15 to 30 f Gm.
Ammon. Aqua	10 to 30 min.	5 to 20 f D.
" Spir.	10 to 30 min.	5 to 20 f D.
" " Arom.	30 to 60 min.	2 to 4 f Gm.
" " Fœtid.	30 to 60 min.	2 to 4 f Gm.
" " Acet. Sol.	2 to 8 fl. dr.	5 to 30 f Gm.
" Benzoas	10 to 20 gr.	50 to 120 cents.
" Bromid.	5 to 20 gr.	25 to 120 cents.
" Carb.	3 to 10 gr.	15 to 60 cents.

Remedies.	Dose in Apothecaries' Weights and Measures.	Dose in Metric Terms.
Ammon. Chlorid.	5 to 30 gr.	0.30 to 2 Gm.
" Iodid.	3 to 10 gr.	15 to 60 cents.
" Phosph.	5 to 20 gr.	30 to 120 cents.
" Picras	¼ to ½ gr.	1 to 3 cents.
" Succin.	1 to 8 gr.	5 to 50 cents.
" Sulph.	20 to 30 gr.	1 to 2 Gm.
" Valer.	2 to 8 gr.	10 to 50 cents.
Ampelopsin (eclectic)	2 to 4 gr.	10 to 25 cents.
Ampelopsis	5 to 30 gr.	0.20 to 2 Gm.
Amygd. Am. Aetherol.	1/16 to ½ min.	0.05 to 0.3 f D.
" " Aqua	1 to 3 fl. dr.	4 to 10 f Gm.
" Oleum	1 to 8 fl. dr.	4 to 30 f Gm.
Amyl. Nitras (see next line)	2 to 5 min.	1 to 3 f D.
" Nitris	2 to 5 min.	1 to 3 f D.
Anemone	1 to 5 gr.	5 to 30 cents.
Angel. Rad	20 to 60 gr.	1 to 4 Gm.
Angustura Cort.	15 to 30 gr.	1 to 2 Gm.
Anilina (and its salts)	½ to 1½ gr.	3 to 10 cents.
Anisum	15 to 30 gr.	1 to 2 Gm.
" Aetherol.	5 to 15 min.	2 to 10 f D.
" Spir.	1 to 2 fl. dr.	4 to 8 f Gm.
Anthemis	10 to 30 gr.	0.60 to 2 Gm.
" Aetherol.	3 to 10 min.	2 to 6 f D.
" Inf.	1 to 3 fl. oz.	30 to 100 f Gm.
Antimon. Oxid.	2 to 3 gr.	10 to 20 cents.
" Pot. Tart. diaphor.	1/32 to ¼ gr.	2 to 15 mills.
" " emet.	1 to 3 gr.	5 to 20 cents.
" Oxysulph.	1 to 3 gr.	5 to 20 cents.
" Sulphurat. alter.	1 to 3 gr.	5 to 20 cents.
" " emet.	5 to 20 gr.	30 to 120 cents.
" Pulvis	3 to 10 gr.	15 to 60 cents.
" Vinum expect.	2 to 3 min.	1 to 5 f D.
" " emet.	½ to 2 fl. dr.	2 to 8 f Gm.
Apiol	4 to 15 gr.	20 to 100 cents.
Apocynin (eclectic)	½ to 1 gr.	2 to 6 cents.
Apocyn. Andros.	8 to 60 gr.	0.50 to 4 Gm.

Remedies.	Dose in Apothecaries' Weights and Measures.	Dose in Metric Terms.
Apocyn. Cannab.	5 to 30 gr.	0.30 to 2 Gm.
Apomorphina (and salts; hypoderm.)	$\frac{1}{60}$ to $\frac{1}{10}$ gr.	1 to 6 mills.
Aqua Acid. Carbol.	1 to 2 fl. dr.	4 to 8 f Gm.
" Ammon. (10 p. c.)	10 to 30 min.	5 to 20 f D.
" Amygd. Amar.	1 to 3 fl. dr.	4 to 12 f Gm.
" Camph.	2 to 8 fl. dr.	8 to 30 f Gm.
" Chlorum	1 to 4 fl. dr.	4 to 15 f Gm.
" Cinnam.	1 to 2 fl. oz.	30 to 60 f Gm.
" Creosotum	1 to 4 fl. dr.	4 to 15 f Gm.
" Laurocer.	$\frac{1}{2}$ to 1 fl. dr.	2 to 4 f Gm.
" Pix	2 to 4 fl. oz.	30 to 120 f Gm.
Aralia Hisp.	$\frac{1}{2}$ to 2 dr.	2 to 8 Gm.
" Nud.	$\frac{1}{2}$ to 2 dr.	2 to 8 Gm.
" Racem.	$\frac{1}{2}$ to 2 dr.	2 to 8 Gm.
" Spin.	$\frac{1}{2}$ to 2 dr.	2 to 8 Gm.
Araroba	5 to 20 gr.	0.30 to 1.50 Gm.
Areca	1 to 4 dr.	4 to 15 Gm.
Argent. Iodid.	1 to 2 gr.	5 to 10 cents.
" Nitr.	$\frac{1}{8}$ to 2 gr.	1 to 10 cents.
" Oxid.	$\frac{1}{4}$ to 2 gr.	1 to 10 cents.
Arnica Flor.	5 to 20 gr.	30 to 120 cents.
" " Tinct.	5 to 30 min.	3 to 20 f D.
" Rad.	5 to 30 gr.	0.30 2 Gm.
" " Ext.	3 to 5 gr.	10 to 30 cents.
" " Tinct.	5 to 30 min.	3 to 20 f D.
Arsenic. Acid.	$\frac{1}{60}$ to $\frac{1}{12}$ gr.	1 to 5 mills.
Arsenios. Acid.	$\frac{1}{60}$ to $\frac{1}{12}$ gr.	1 to 5 mills.
Arsen. Brom.	$\frac{1}{60}$ to $\frac{1}{20}$ gr.	1 to 3 mills.
" Iodid.	$\frac{1}{60}$ to $\frac{1}{20}$ gr.	1 to 3 mills.
" Sol. Hydrochlor.	2 to 8 min.	1 to 5 f D.
Artemisia Abrot.	10 to 20 gr.	60 to 120 cents.
" Vulg.	15 to 60 gr.	1 to 4 Gm.
Arum	15 to 30 gr.	1 to 2 Gm.
Asafœtida	5 to 10 gr.	20 to 60 cents.
" Mixt.	4 to 8 fl. dr.	15 to 30 f Gm.

Remedies.	Dose in Apothecaries' Weights and Measures.	Dose in Metric Terms.
Asafœtida Pil.	1 to 3 pills	1 to 3 pills.
" Tinct.	$\frac{1}{2}$ to 1 fl. dr.	2 to 4 f Gm.
Asarum	15 to 30 gr.	1 to 2 Gm.
Asclep. Incarn.	8 to 40 gr.	50 to 250 cents.
" Syriac.	8 to 60 gr.	0.50 to 4 Gm.
" Tuber.	30 to 60 gr.	2 to 4 Gm.
Asclepin (eclectic)	2 to 4 gr.	10 to 20 cents.
Atropina (and salts)	$\frac{1}{100}$ to $\frac{1}{30}$ gr.	0.5 to 2 mills.
Atropin (eclectic resinoid)	$\frac{1}{24}$ to $\frac{1}{12}$ gr.	2 to 5 mills.
Aurant Am. Cort.	39 to 60 gr.	2 to 4 Gm.
" " " Tinct.	1 to 2 fl. dr.	4 to 8 f Gm.
" Dulc. Cort.	30 to 60 gr.	2 to 4 Gm.
" " " Tinct.	1 to 2 fl. dr.	4 to 8 f Gm.
Aur. Sod. Chlorid.	$\frac{1}{32}$ to $\frac{1}{15}$ gr.	2 to 4 mills.
Ava Kava	30 to 60 gr.	2 to 4 Gm.
Azedarach	1 to 2 dr.	4 to 8 Gm.
Bals. Dipteroc.	15 to 30 min.	1 to 2 f Gm.
" Gurjun.	15 to 30 min.	1 to 2 f Gm.
" Peruv.	15 to 30 min.	1 to 2 f Gm.
" Tolut.	15 to 30 gr.	1 to 2 Gm.
Baptisia	5 to 20 gr.	0.20 to 1.20 Gm.
Baptisin (eclectic)	1 to 3 gr.	5 to 20 cents
Bar. Chlorid.	$\frac{1}{8}$ to 1 gr.	2 to 6 cents
Barosma Fol.	15 to 30 gr.	1 to 2 Gm.
" " Tinct.	1 to 2 fl. dr.	4 to 8 f Gm.
Barosmin (eclectic)	2 to 3 gr.	10 to 20 cents.
Beheerina Sulph.	1 to 10 gr.	5 to 60 cents.
Bela Fruct.	1 to 4 dr.	5 to 15 Gm.
Bellad. Fol.	1 to 10 gr.	5 to 60 cents.
" " Extr.	$\frac{1}{4}$ to $\frac{3}{4}$ gr.	15 to 50 mills.
" " " Alcoh.	$\frac{1}{3}$ to $\frac{1}{2}$ gr.	2 to 3 cents.
" " Tinct.	15 to 30 min.	1 to 2 f Gm.
" Rad.	1 to 5 gr.	5 to 30 cents.
" " Extr.	$\frac{1}{8}$ to $\frac{1}{4}$ gr.	8 to 15 mills.
" " Tinct.	10 to 20 min.	6 to 12 f D.
Benzoin Odorif.	30 to 60 gr.	2 to 4 Gm.

Remedies.	Dose in Apothecaries' Weights and Measures.	Dose in Metric Terms.
Benzoe Tinct.	15 to 30 min.	1 to 2 f Gm.
" " Comp.	$\frac{1}{2}$ to 2 fl. dr.	2 to 8 f Gm.
Berberina (and salts)	1 to 15 gr.	5 to 100 cents.
Berberis Aquif.	15 to 30 gr.	1 to 2 Gm.
" Vulg.	$\frac{1}{2}$ to 2 dr.	2 to 8 Gm.
Bismuth. Ammon. Citr.	2 to 5 gr.	10 to 30 cents.
" Boras	1 to 5 gr.	5 to 30 cents.
" Citras	1 to 3 gr.	5 to 20 cents.
" Subcarb.	8 to 60 gr.	0.50 to 4 Gm.
" Subnitr.	5 to 15 gr.	30 to 100 cents.
" Tannas	10 to 30 gr.	60 cts. to 2 Gm.
" Valer.	$\frac{1}{2}$ to 3 gr.	3 to 20 cents
Boldus Fol.	10 to 20 gr.	60 to 120 cents.
Borax	5 to 30 gr.	30 cts. to 2 Gm.
Brayera Flor.	3 to 4 dr.	10 to 15 Gm.
" " Infus.	4 to 8 fl. oz.	100 to 250 f Gm.
Bromoformum	2 to 10 min.	1 to 6 f D.
Bromum	$\frac{1}{2}$ to 2 gr.	3 to 10 cents.
Brucina and its salts	$\frac{1}{30}$ to $\frac{1}{15}$ gr.	2 to 4 mills.
Bryonia Rad.	15 to 60 gr.	1 to 4 Gm.
" " Tinct.	1 to 4 fl. dr.	4 to 15 f Gm.
" " Recens Tinct.	$\frac{1}{2}$ to 1 fl. dr.	2 to 4 f Gm.
Buchu	15 to 30 gr.	1 to 2 Gm.
Cactus Grandiflor.	2 to 5 gr.	10 to 30 cents.
Cadm. Sulph.	$\frac{1}{6}$ to 1 gr.	1 to 6 cents.
Caffea	15 to 60 gr.	1 to 4 Gm.
" Tosta	$\frac{1}{2}$ to 2 dr.	2 to 8 Gm.
Coffeina (and salts)	1 to 5 gr.	5 to 30 cents.
Calabarina	$\frac{1}{64}$ to $\frac{1}{18}$ gr.	1 to 15 mills.
Calamus	15 to 60 gr.	1 to 4 Gm.
" Tinct.	1 to 4 fl. dr.	4 to 15 f Gm.
Calc. Brom.	5 to 30 gr.	30 cts. to 2 Gm.
" Carb.	10 to 40 gr.	60 to 250 cents.
" Chlor.	10 to 20 gr.	60 to 120 cents.
" Hypophosph.	10 to 30 gr.	60 cts. to 2 Gm.
" Iodid	1 to 5 gr.	6 to 30 cents.

Remedies.	Dose in Apothecaries' Weights and Measures.	Dose in Metric Terms.
Calc. Phosph.	10 to 30 gr.	60 cts. to 2 Gm.
" Sacch. (Syr.)	15 to 30 min.	1 to 2 f Gm.
" Sulphuret.	$\frac{1}{20}$ to 1 gr.	3 to 60 mills.
Calendula	$\frac{1}{2}$ to 2 dr.	2 to 8 Gm.
" Extr.	1 to 3 gr.	5 to 20 cents.
" Tinct.	1 to 2 fl. dr.	2 to 4 f Gm.
Calumba	15 to 30 gr.	1 to 2 Gm.
" Infus.	1 to 2 fl. oz.	30 to 60 f Gm.
" Tinct.	1 to 4 fl. dr.	4 to 15 f Gm.
Calx Chlorata	3 to 6 gr.	15 to 40 cents.
" Sulphurata	$\frac{1}{20}$ to 1 gr.	3 mills to 5 cts.
" Syrupus	15 to 30 min.	1 to 2 f Gm.
Camphora	1 to 10 gr.	5 to 60 cents.
" Aqua	2 to 8 fl. dr.	10 to 30 f Gm.
" Spir.	5 to 60 min.	3 to 40 f D.
" Monobrom.	2 to 6 gr.	10 to 40 cents.
Canella	5 to 10 gr.	25 to 60 cents.
Cannabin (eclectic).	$\frac{1}{8}$ to $\frac{1}{4}$ gr.	8 to 16 mills.
Cannabis Amer.	5 to 20 gr.	25 to 130 cents.
" " Tinct.	5 to 20 min.	3 to 15 f D.
" Ind.	5 to 20 gr.	30 to 130 cents.
" " Extr.	$\frac{1}{4}$ to $\frac{1}{2}$ gr.	15 to 30 cents.
" " Tinct.	5 to 15 min.	3 to 10 f D.
Cantharides	1 to 2 gr.	5 to 10 cents.
" Fl. Extr.	1 to 2 min.	0.5 to 1 f D.
" Tinct.	5 to 20 min.	3 to 15 f D.
Capsicum	1 to 5 gr.	5 to 30 cents.
" Oleoresina	$\frac{1}{8}$ to 1 gr.	8 mills to 6 cts.
" Tinct.	10 to 20 min.	6 to 40 f D.
Carbo Ligni	1 to 4 dr.	4 to 15 Gm.
Carbon. Bisulphid.	$\frac{1}{2}$ to 1$\frac{1}{2}$ min.	0.3 to 1 f D.
Cardamomum	10 to 30 gr.	50 cts. to 2 Gm.
" Oleoresina	$\frac{1}{4}$ to 2 gr.	1 to 10 cents.
" Tinct.	$\frac{1}{2}$ to 2 fl. dr.	2 to 10 f Gm.
Carduus Bened.	10 to 60 gr.	50 cts. to 5 Gm.
Carota Fruct.	30 to 60 gr.	2 to 5 Gm.

Remedies.	Dose in Apothecaries' Weights and Measures.	Dose in Metric Terms.
Carthamus	1 to 2 dr.	5 to 10 Gm.
Carum	30 to 60 gr.	2 to 5 Gm.
Caryophyllus	5 to 10 gr.	20 to 60 cents.
Cascara Sagrada, lax.	10 to 20 gr.	50 cts. to 1 Gm.
" " cath.	15 to 60 gr.	1 to 5 Gm.
Cascarilla	15 to 30 gr.	1 to 2 Gm.
" Extr.	1 to 5 gr.	5 to 30 cents.
" Tinct.	½ to 2 fl. dr.	2 to 10 f Gm.
Cassia Fistula Pulpa lax.	1 to 2 dr.	5 to 10 Gm.
" " " purg.	1 to 2 oz.	30 to 60 Gm.
" Maryland	½ to 2 dr.	2 to 5 Gm.
Castanea	15 to 60 gr.	1 to 5 Gm.
Castoreum	15 to 20 gr.	1 to 1.50 Gm.
" Tinct.	½ to 2 fl. dr.	2 to 10 f Gm.
Cataria	½ to 1 dr.	2 to 5 Gm.
Catechu	5 to 30 gr.	25 cts. to 2 Gm.
" Tinct.	½ to 2 fl. dr.	2 to 10 f Gm.
Caulophyllin (eclectic)	1 to 5 gr.	5 to 30 cents.
Caulophyllum	5 to 30 gr.	20 to 200 cents.
Cera, Alba and Flava	15 to 30 gr.	1 to 2 Gm.
Cerasein (eclectic)	2 to 10 gr.	10 to 60 cents.
Cereus Grandiflor.	2 to 5 gr.	10 to 30 cents.
Cer. Nitr.	1 to 3 gr.	5 to 20 cents.
" Oxal.	1 to 3 gr.	5 to 20 cents.
Cetraria Decoct.	1 to 2 fl. oz.	30 to 60 f Gm.
Chamælirium	15 to 45 gr.	1 to 3 Gm.
Chelidonium	15 to 30 gr.	1 to 2 Gm.
Chelone	1 to 2 dr.	4 to 10 Gm.
Chelonin (eclectic)	1 to ·2 gr.	5 to 15 cents.
Chenopod. Aetherol.	3 to 8 min.	2 to 5 f D.
" Sem.	15 to 45 gr.	1 to 3 Gm.
Chian Turpentine	15 to 70 gr.	1 to 5 Gm.
Chimaphila	30 to 70 gr.	2 to 5 Gm.
" Tinct.	15 to 60 min.	1 to 4 f Gm.
Chinoidinum	3 to 30 gr.	15 cents to 2 Gm.
" Tinct.	1 to 2 fl. dr.	5 to 10 f Gm.

Remedies.	Dose in Apothecaries' Weights and Measures.	Dose in Metric Terms.
Chionanthus	30 to 70 gr.	2 to 5 Gm.
Chionanthin (eclectic)	1 to 3 gr.	5 to 20 cents.
Chirata	30 to 70 gr.	2 to 5 Gm.
" Tinct.	15 to 60 min.	1 to 4 f Gm.
Chloral Hydrate	5 to 30 gr.	25 cents to 2 Gm.
" Croton	1 to 10 gr.	5 to 60 cents.
Chlorodyne	5 to 20 min.	3 to 15 f D.
Chloroformum	2 to 40 min.	1 f D. to 3 f Gm.
" Aqua	1 to 4 fl. dr.	5 to 15 f Gm.
Chondodendrum	½ to 2 dr.	2 to 10 Gm.
Chondrus	2 to 4 dr.	5 to 15 Gm.
Cicuta	3 to 8 gr.	15 to 50 cents.
Cimicifuga	15 to 60 gr.	1 to 4 Gm.
" Tinct.	½ to 2 fl. dr.	2 to 8 f Gm.
Cimicifugin (eclectic)	1 to 3 gr.	5 to 20 cents.
Cinchona	15 to 60 gr.	1 to 4 Gm.
" Extr.	5 to 30 gr.	15 cents to 2 Gm.
" Tinct.	½ to 4 fl. dr.	2 to 15 f Gm.
" " Comp.	½ to 4 fl. dr.	2 to 15 f Gm.
Cinchonidina (and salts)	1 to 20 gr.	5 to 120 cents.
Cinchonina (and salts)	1 to 20 gr.	5 to 120 cents.
Cinnamomum	10 to 20 gr.	60 to 120 cents.
" Aetherol	1 to 3 min.	1 to 2 f D.
" Tinct.	1 to 3 fl. dr.	4 to 12 f Gm.
Coca	½ to 2 dr.	2 to 8 Gm.
Coccionella	⅓ to 1 gr.	2 to 5 cents.
" Tinct.	15 to 60 min.	1 to 4 Gm.
Cocculus	1 to 3 gr.	5 to 20 cents.
Coccus	⅓ to 1 gr.	2 to 5 cents.
Codeina (and salts)	½ to 3 gr.	2 to 20 cents.
Colch. Rad.	2 to 8 gr.	10 to 50 cents.
" " Extr. Acet.	1 to 2 gr.	5 to 10 cents.
" " Tinct.	5 to 20 min.	2 to 12 f D.
" " Vin.	10 to 30 min.	6 to 20 f D.
" Sem.	2 to 10 gr.	10 to 60 cents.
" " Tinct.	15 to 60 min.	1 to 4 f Gm.

Remedies.	Dose in Apothecaries' Weights and Measures.	Dose in Metric Terms.
Colch. Sem. Vin.	15 to 60 min.	1 to 4 f Gm.
Collinsonia	5 to 30 gr.	25 cents to 2 Gm.
Collinsonin (eclectic)	2 to 4 gr.	10 to 25 cents.
Colocynthis	5 to 10 gr.	25 to 60 cents.
" Extr.	5 to 20 gr.	25 to 120 cents.
" " Comp.	5 to 30 gr.	25 cts. to 2 Gm.
" Tinct.	½ to 2 fl. dr.	2 to 8 f Gm.
Comptonia	30 to 60 gr.	2 to 4 Gm.
Condurango	30 to 60 gr.	2 to 4 Gm.
Conf. Arom.	8 to 60 gr.	50 cts. to 4 Gm.
" Aurant.	1 to 4 dr.	5 to 15 Gm.
" Cinch. Co.	1 to 4 dr.	5 to 15 Gm.
" Cubeba Co.	1 to 2 dr.	5 to 10 Gm.
" Opii	5 to 30 gr.	25 cts. to 2 Gm.
" Piper	1 to 2 dr.	4 to 8 Gm.
" Rosa	30 to 60 gr.	2 to 4 Gm.
" Scammon	8 to 30 gr.	50 cts. to 2 Gm.
" Senna	1 to 2 dr.	4 to 8 Gm.
" Sulphur	1 to 3 dr.	4 to 12 Gm.
Coniina (and salts)	$\frac{1}{80}$ to $\frac{1}{30}$ gr.	1 to 2 mills.
Conium Folia	3 to 8 gr.	15 to 50 cents.
" " Extr.	1 to 3 gr.	3 to 20 cents.
" " " Alcola.	1 to 2 gr.	5 to 10 cents.
" " Tinct.	½ to 2 fl. dr.	2 to 8 f Gm.
" Sem.	½ to 4 gr.	2 to 25 cents.
" " Extr.	1 to 2 gr.	5 to 10 cents.
" " Tinct.	15 to 60 min.	1 to 4 f Gm.
Convallaria	30 to 60 gr.	2 to 4 Gm.
Copaiba Balsam	10 to 40 min.	6 to 25 f D.
" Aetherol	10 to 15 min.	5 to 10 f D.
" Pilula	15 to 60 gr.	1 to 4 Gm.
" Resina	2 to 10 gr.	10 to 60 cents.
Coptis	30 to 60 gr.	2 to 4 Gm.
Corallorrhiza	15 to 30 gr.	1 to 2 Gm.
Coriandrum	15 to 60 gr.	1 to 4 Gm.
Cornin (eclectic)	2 to 4 gr.	10 to 25 cents.

Remedies.	Dose in Apothecaries' Weights and Measures.	Dose in Metric Terms.
Cornus Circin.	30 to 60 gr.	2 to 4 Gm.
" Florida	30 to 60 gr.	2 to 4 Gm.
" Sericea	30 to 60 gr.	2 to 4 Gm.
Corydalin (eclectic)	2 to 4 gr.	10 to 25 cents.
Corydalis	10 to 40 gr.	60 cts. to 250 Gm
Coto Bark	$\frac{1}{2}$ to 8 gr.	3 to 50 cents.
Cotoinum	$\frac{1}{10}$ to $\frac{1}{2}$ gr.	5 to 25 mills.
Cotula	1 to 2 dr.	4 to 8 Gm.
Cratæva Marmelos	1 to 2 dr.	4 to 8 Gm.
Creosotum	1 to 3 min.	1 to 2 f D.
Creta Præp.	10 to 40 gr.	60 cts. to 2.50 Gm
" Pulv. Arom.	30 to 60 gr.	2 to 4 Gm.
Crocus	10 to 20 gr.	60 cts. to 1.20 Gm
" Tinct.	1 to 3 fl. dr.	4 to 15 f Gm.
Croton Chloral Hydrate	1 to 10 gr.	5 to 60 cents.
Cubeba	10 gr. to 3 dr.	60 cts. to 12 Gm.
" Aetherol	5 to 20 min.	3 to 12 f D.
" Conf.	30 to 60 gr.	2 to 4 Gm.
" Fluidextr.	10 to 40 min.	6 to 25 f D.
" Oleores.	5 to 30 gr.	25 cts. to 2 Gm.
" Tinct.	$\frac{1}{2}$ to 2 fl. dr.	2 to 8 f Gm.
Cupr. Acet.	$\frac{1}{4}$ to 1 gr.	1 to 5 cents.
" Ammon.	$\frac{1}{4}$ to $\frac{1}{2}$ gr.	1 to 3 cents.
" Sulph. Emet.	2 to 10 gr.	10 to 60 cents.
" " Astring.	$\frac{1}{4}$ to $\frac{1}{2}$ gr.	1 to 3 cents.
Curare	$\frac{1}{12}$ to $\frac{1}{2}$ gr.	5 to 30 mills.
Curarina	$\frac{1}{60}$ to $\frac{1}{20}$ gr.	1 to 3 mills.
Curcuma	10 to 40 gr.	50 cts. to 2.50 Gm
Cusparia	10 to 40 gr.	50 cts. to 2.50 Gm
Cyclaminum	$\frac{1}{2}$ to 1$\frac{1}{2}$ gr.	3 to 10 cents.
Cypripedin (eclectic)	$\frac{1}{2}$ to 3 gr.	3 to 20 cents.
Cypripedium	15 to 30 gr.	1 to 2 Gm.
" Tinct.	1 to 2 fl. dr.	4 to 8 f Gm.
Damiana	1 to 2 dr.	4 to 8 Gm.
Daturina (and salts)	$\frac{1}{120}$ to $\frac{1}{30}$ gr.	0.50 to 2 mills.
Decoct. Aloës. Co.	$\frac{1}{2}$ to 2 fl. oz.	15 to 60 f Gm.

Remedies.	Dose in Apothecaries' Weights and Measures.	Dose in Metric Terms.
Decoct. Cetrar.	1 to 2 fl. oz.	30 to 60 f Gm.
" Chimaph.	3 to 6 fl. oz.	100 to 200 f Gm.
" Cinch. Fl.	1 to 2 fl. oz.	30 to 60 f Gm.
" " Rubr.	1 to 2 fl. oz.	30 to 60 f Gm.
" Cornus Flor.	1 to 2 fl. oz.	30 to 60 f Gm.
" Dulcamara	1 to 2 fl. oz.	30 to 60 f Gm.
" Geranium	1 to 2 fl. oz.	30 to 60 f Gm.
" Granat.	1 to 2 fl. oz. ˙	30 to 60 f Gm.
" Hæmatox.	1 to 2 fl. oz.	30 to 60 f Gm.
" Hordeum	4 to 8 fl. oz.	100 to 250 f Gm.
" Pareira	1 to 2 fl. oz.	30 to 60 f Gm.
" Quercus	1 to 2 fl. oz.	30 to 60 f Gm.
" Sarsap. Co.	2 to 6 fl. oz.	60 to 200 f Gm.
" Senega	1 to 2 fl. oz.	30 to 60 f Gm.
" Ulmus	4 to. 6 fl. oz.	100 to 200 f Gm.
" Uva Ursi	1 to 2 fl. oz.	30 to 60 f Gm.
Delphinium	1 to 3 gr.	5 to 20 cents.
Digitalin (eclectic), see below.	$\frac{1}{8}$ to $\frac{1}{2}$ gr.	8 to 30 mills.
Digitalinum U. S. P. 1870	$\frac{1}{60}$ to $\frac{1}{30}$ gr.	1 to 2 mills.
Digitalis Fol.	$\frac{1}{2}$ to 2 gr.	3 to 12 cents.
" " Extr.	$\frac{1}{8}$ to $\frac{1}{2}$ gr.	8 to 30 mills.
" " Inf.	1 to 4 fl. dr.	5 to 15 f Gm.
" " Tinct.	10 to 20 min.	6 to 12 f D.
" Rad.	$\frac{1}{2}$ to 1 gr.	3 to 6 cents.
Dioscorea	10 to 30 gr.	50 cts. to 2 Gm.
Dioscorein (eclectic)	$\frac{1}{2}$ to 4 gr.	3 to 25 cents.
Diospyros	$\frac{1}{2}$ to 3 dr.	3 to 20 cents.
Dipterix Odor	5 to 10 min.	2 to 6 f D.
Dita	1 to 4 dr.	4 to 15 Gm.
Ditainum	2 to 10 gr.	10 to 60 cents.
Dracontium	10 to 60 gr.	50 cts. to 4 Gm.
Drosera	5 to 10 gr.	25 to 60 cents.
Duboisina (and salts)	$\frac{1}{120}$ to $\frac{1}{60}$ gr.	0.50 to 1 mill.
Dulcamara	30 to 60 gr.	2 to 4 Gm.
" Extr.	5 to 10 gr.	25 to 60 cents.
" Infus.	1 to 2 fl. oz.	30 to 60 f Gm.

Remedies.	Dose in Apothecaries' Weights and Measures.	Dose in Metric Terms.
Elaterinum Album (cryst.)	$\frac{1}{80}$ to $\frac{1}{16}$ gr.	0.80 to 4 mills.
Elaterium Viride	$\frac{1}{32}$ to $\frac{1}{8}$ gr.	2 to 8 mills.
Emetina (and salts) emet.	$\frac{1}{120}$ to $\frac{1}{30}$ gr.	0.50 to 2 mills.
" " diaphor.	$\frac{1}{8}$ to $\frac{1}{4}$ gr.	8 to 15 mills.
Emulsio Hydrocyan.	$\frac{1}{2}$ to 1 fl. dr.	2 to 4 f Gm.
Epigæa	30 to 60 gr.	2 to 4 Gm.
Ergota	15 to 60 gr.	1 to 4 Gm.
" Extr.	2 to 12 gr.	10 to 70 cents.
" Liquor.	10 to 30 min.	6 to 20 f D.
" Tinct.	15 to 60 min.	1 to 4 f Gm.
" Vin.	1 to 3 fl. dr.	4 to 12 f Gm.
Ergotin	2 to 10 gr.	5 to 60 cents.
Erigeron	30 to 60 gr.	2 to 4 Gm.
" Aetherol	3 to 8 min.	1 to 5 f D.
Eriodictyon	15 to 60 gr.	1 to 4 Gm.
Erythroxylon	$\frac{1}{2}$ to 2 dr.	2 to 8 Gm.
" Tinct.	$\frac{1}{2}$ to 2 fl. oz.	15 to 60 f Gm.
Erythroxylin (eclectic)	$\frac{1}{2}$ to 3 gr.	2 to 20 cents.
Eserina (and salts)	$\frac{1}{64}$ to $\frac{1}{20}$ gr.	1 to 3 mills.
Eucalyptus	15 to 60 gr.	1 to 4 Gm.
" Aetherol.	5 to 30 min.	2 to 20 f D.
" Tinct.	1 to 3 fl. dr.	4 to 12 f Gm.
Euonymin (eclectic)	$\frac{1}{2}$ to 3 gr.	2 to 20 cents.
Euonymus	1 to 2 dr.	4 to 8 Gm.
" Extr.	1 to 10 gr.	5 to 60 cents.
Eupatorin (eclectic)	1 to 3 gr.	5 to 20 cents.
Eupatorium	1 to 2 dr.	4 to 8 Gm.
Euphorbia Coroll	3 to 10 gr.	15 to 60 cents.
" Ipecac.	10 to 15 gr.	50 to 100 cents.
Eupurpurin (eclectic)	1 to 3 gr.	5 to 20 cents.
Extracta Fluida *		
Extr. Aconit. Fol.	$\frac{1}{4}$ to 1 gr.	1 to 6 cents.
" " Rad.	$\frac{1}{6}$ to $\frac{1}{4}$ gr.	1 to 2 cents.
" Aloē	$\frac{1}{2}$ to 3 gr.	2 to 20 cents.
" Anthemis	2 to 10 gr.	10 to 60 cents.
" Arnica Flor.	3 to 10 gr.	15 to 60 cents.

* Fluid extracts will be found under letter F.

Remedies.	Dose in Apothecaries' Weights and Measures.	Dose in Metric Terms.
Extr. Arnica Rad.	3 to 10 gr.	15 to 60 cents.
" Bellad.	¼ to ¾ gr.	1 to 4 cents.
" " Fol. Alc.	⅙ to ½ gr.	1 to 3 cents.
" " Rad.	⅛ to ¼ gr.	8 to 15 mills.
" Calumba	2 to 10 gr.	10 to 60 cents.
" Cannab. Amer.	¼ to ¾ gr.	1 to 4 cents.
" " Ind.	⅙ to ½ gr.	1 to 3 cents.
" Carnis	10 to 60 gr.	1 to 4 Gm.
" Cinch.	10 to 30 gr.	1 to 2 Gm.
" Colch. Acet.	½ to 2 gr.	2 to 10 cents.
" Coloc.	2 to 5 gr.	10 to 30 cents.
" " Co.	2 to 5 gr.	10 to 30 cents.
" Conium	1 to 4 gr.	5 to 25 cents.
" " Fol. Alc.	1 to 2 gr.	5 to 10 cents.
" " Fruct.	½ to 1½ gr.	2 to 10 cents.
" Digit. Fol.	¼ to ½ gr.	1 to 3 cents.
" " Rad.	⅛ to ⅓ gr.	8 to 20 mills.
" Dulcam.	5 to 10 gr.	25 to 60 cents.
" Ergota	2 to 10 gr.	10 to 60 cents.
" Ferr. Pom.	3 to 10 gr.	15 to 60 cents.
" Gent.	10 to 30 gr.	50 cts. to 2 Gm.
" Glycyrrh.	30 to 60 gr.	2 to 4 Gm.
" " Purif.	20 to 60 gr.	1 to 4 Gm.
" Gramen	1 to 3 dr.	4 to 12 Gm.
" Hæmat	10 to 30 gr.	1 to 2 Gm.
" Helleb.	3 to 10 gr.	15 to 60 cents.
" Humul.	3 to 20 gr.	15 to 120 cents.
" Hyoscyam.	1 to 4 gr.	5 to 25 cents.
" Hyoscyam. Fol. Alc.	1 to 2 gr.	5 to 10 cents.
" " " Sem.	1 to 2 gr.	5 to 10 cents.
" Ignatia	¼ to 1½ gr.	1 to 10 cents.
" Jalapa	5 to 15 gr.	25 cts. to 1 Gm.
" " Alc.	3 to 6 gr.	15 to 35 cents.
" Juglans	20 to 30 gr.	1 to 2 Gm.
" Krameria	5 to 20 gr.	25 to 120 cents.
" Lactuca	5 to 10 gr.	25 to 60 cents.

Remedies.	Dose in Apothecaries' Weights and Measures.	Dose in Metric Terms.
Extr. Lupulinum	5 to 10 gr.	25 to 60 cents.
" Maltum	2 to 4 dr.	5 to 15 Gm.
" Nux Vom.	¼ to 1 gr.	1 to 6 cents.
" Opium	¼ to ¾ gr.	1 to 4 cents.
" Papaver	2 to 5 gr.	10 to 30 cents.
" Pareira	10 to 20 gr.	1 to 1.20 Gm.
" Physostigma	1/13 to ½ gr.	4 to 10 mill.
" Podophyll.	5 to 15 gr.	25 cts. to 1 Gm.
" Quassia	3 to 10 gr.	15 to 60 cents.
" Rheum	5 to 15 gr.	25 cts. to 1 Gm.
" Senega	1 to 3 gr.	5 to 20 cents.
" Stramon Fol.	½ to 1 gr.	2 to 6 cents.
" " Sem.	¼ to ½ gr.	1 to 3 cents.
" Tarax.	15 to 40 gr.	1 to 2.50 Gm.
" Tritic.	1 to 3 dr.	4 to 12 Gm.
" Valer.	15 to 30 gr.	1 to 2 Gm.
Fel. Bovinum Purif.	3 to 10 gr.	15 to 60 cents.
Ferr. Arsen.	1/20 to ½ gr.	3 to 30 mills.
" Benz.	1 to 5 gr.	5 to 30 cents.
" Brom.	1 to 5 gr.	5 to 30 cents.
" " Syrup	15 to 60 min.	1 to 4 f Gm.
" Carb. Sacch.	5 to 20 gr.	25 to 120 cents.
" Chlorid.	1 to 5 gr.	5 to 30 cents.
" " Sol.	8 to 30 min.	5 to 20 f D.
" " " Spir.	8 to 30 min.	5 to 20 f D.
" Citr.	5 to 10 gr.	25 to 60 cents.
" Ammon. Citr.	5 to 10 gr.	25 to 60 cents.
" " Sulph.	5 to 15 gr.	25 cts. to 1 Gm.
" " Tartr.	8 to 30 gr.	50 cts. to 2 Gm.
" Bism. Citr.	15 to 60 gr.	1 to 4 Gm.
" Cinchonid Citr.	5 to 15 gr.	25 cts. to 1 Gm.
" Mangan. Carb. Sacch.	5 to 20 gr.	25 to 120 cents.
" Pot. Tartr.	8 to 30 gr.	50 cts. to 2 Gm.
" Quin. Citr.	5 to 10 gr.	25 to 60 cents.
" Quinid. Citr.	5 to 10 gr.	25 to 60 cents.
" Strychn. Citr.	1 to 5 gr.	5 to 30 cents.

Remedies.	Dose in Apothecaries' Weights and Measures.	Dose in Metric Terms.
Ferr. Quin. Strychn. Citr.	3 to 10 gr.	15 to 60 cents.
" Ferrocyanid.	3 to 5 gr.	15 to 30 cents.
" Hypophosph.	5 to 10 gr.	25 to 60 cents.
" Iodidum	1 to 5 gr.	5 to 30 cents.
" Lactas	1 to 2 gr.	5 to 12 cents.
" Oxalas	2 to 3 gr.	10 to 20 cents.
" Oxid. Hydrat. in Aqua	½ to 2 oz.	15 to 60 Gm.
" " Magnet.	5 to 10 gr.	25 to 60 cents.
" Phosph. Alb.	1 to 5 gr.	5 to 30 cents.
" " Cœrul.	5 to 10 gr.	25 to 60 cents.
" Pyrophosph.	2 to 5 gr.	10 to 30 cents.
" Subcarb.	5 to 30 gr.	25 cts. to 2 Gm.
" Sulph.	1 to 3 gr.	5 to 20 cents.
" " Exsicc.	½ to 2 gr.	2 to 12 cents.
" Tannas	5 to 30 gr.	25 cts. to 2 Gm.
" Valerian.	½ to 2 gr.	2 to 12 cents.
Ferrum Dialys.	1 to 15 min.	1 to 10 f D.
" Reduct.	1 to 5 gr.	5 to 30 cents.
Filix Mas.	1 to 3 dr.	4 to 12 Gm.
" " Extr.	10 to 30 gr.	60 cts. to 2 Gm.
" " Oleoresina	15 to 30 gr.	1 to 2 Gm.
Fluidextr. Absinth	10 to 40 min.	6 to 25 f D.
" Achillea.	30 to 60 min.	2 to 4 f Gm.
" Aconit Fol.	1 to 5 min.	1 to 4 f D.
" " Rad.	1 to 3 min.	1 to 2 f D.
" Aescul. Glabr.	2 to 8 fl. dr.	5 to 30 f Gm.
" " Hippocast.	10 to 40 min.	6 to 25 f D.
" Agrimonia	½ to 2 fl. dr.	2 to 8 f Gm.
" Ailanthus	10 to 30 min.	6 to 20 f D.
" Aletris	20 to 40 min.	10 to 25 f D.
" Alnus	20 to 60 min.	10 to 40 f D.
" Aloë Soc.	2 to 5 min.	1 to 4 f D.
" Alstonia	1 to 4 fl. dr.	4 to 15 f Gm.
" Ampelopsis	20 to 40 min.	10 to 25 f D.
" Anemone	2 to 5 min.	1 to 4 f D.
" Angustura	15 to 40 min.	10 to 25 f D.

Remedies.	Dose in Apothecaries' Weights and Measures.	Dose in Metric Terms.
Fluidextr. Angelica Rad.	30 to 60 min.	2 to 4 f Gm.
" " Sem.	30 to 60 min.	2 to 4 f Gm.
" Anisum	20 to 30 min.	1 to 2 f Gm.
" Anthemis	30 to 60 min.	2 to 4 f Gm.
" Apocyn. Andros.	10 to 60 min.	6 to 40 f.D.
" " Cannab.	3 to 10 min.	2 to 6 f D.
" Aral. Hispida	30 to 60 min.	2 to 4 f Gm.
" " Nudic.	30 to 60 min.	2 to 4 f Gm.
" " Racem.	2 to 4 fl. dr.	4 to 15 f Gm.
" " Spin.	1 to 2 fl. dr.	4 to 8 f Gm.
" Areca	1 to 4 fl. dr.	5 to 15 f Gm.
" Arnica Flor.	5 to 20 min.	3 to 15 f D.
" " Rad.	5 to 20 min.	3 to 15 f D.
" Arom. (Pulv. Arom.)	30 to 60 min.	2 to 4 f Gm.
" Artemis Abrot.	10 to 20 min.	6 to 15 f D.
" Artemis Vulg.	10 to 60 min.	6 to 40 f D.
" Arum	10 to 30 min.	6 to 20 f D.
" Asarum	1 to 2 fl. dr.	4 to 8 f Gm.
" Asclep. Incarn.	10 to 40 min.	6 to 25 f D.
" " Syr.	10 to 40 min.	6 to 25 f D.
" " Tuber.	½ to 2 fl. dr.	2 to 8 f Gm.
" Aspidium	2 to 4 fl. dr.	8 to 15 f Gm.
" Aurant. Cort.	½ to 2 fl. dr.	2 to 8 f Gm.
" " Fruct.	30 to 60 min.	2 to 4 f Gm.
" Ava Kava	30 to 60 min.	2 to 4 f Gm.
" Azedarach	1 to 2 fl. dr.	4 to 8 f Gm.
" Baptisia	5 to 20 min.	3 to 12 f D.
" Barosma	½ to 2 fl. dr.	2 to 8 f Gm.
" Bellad Fol.	3 to 5 min.	2 to 4 f D.
" " Rad.	1 to 3 min.	1 to 2 f D.
" Benzoin Odorif.	30 to 60 min.	1 to 2 f Gm.
" Berberis Aquifol.	10 to 30 min.	6 to 20 f D.
" " Vulg.	1 to 2 fl. dr.	4 to 8 f Gm.
" Boldus	1 to 10 min.	1 to 6 f D.
" Brayera	2 to 4 fl. dr.	8 to 15 f Gm.

Remedies.	Dose in Apothecaries' Weights and Measures.	Dose in Metric Terms.
Fluidextr. Bryonia	10 to 60 min.	6 to 40 f D.
" Buchu	½ to 2 fl. dr.	2 to 8 f Gm.
" " Comp.	1 to 2 fl. dr.	4 to 8 f Gm.
" Cact. Grandifl.	2 to 5 min.	1 to 4 f D.
" Caffea	2 to 4 fl. dr.	8 to 15 f Gm.
" Calamus	20 to 60 min.	10 to 40 f D.
" Calendula	30 to 60 min.	2 to 4 f Gm.
" Calumba	20 to 60 min.	1 to 4 f Gm.
" Canella	30 to 60 min.	2 to 4 f Gm.
" Cannab. Amer.	2 to 10 min.	1 to 6 f D.
" Cannab. Ind.	2 to 5 min.	1 to 4 f D.
" Cantharides	1 to 3 min.	1 to 2 f D.
" Capsicum	3 to 10 min.	2 to 6 f D.
" Cardam.	30 to 60 min.	2 to 4 f Gm.
" " Co.	30 to 60 min.	2 to 4 f Gm.
" Carduus Bened.	15 to 60 min.	1 to 4 f Gm.
" Carota	30 to 60 min.	2 to 4 f Gm.
" Carthamus	1 to 2 fl. dr.	4 to 8 f Gm.
" Carum	30 to 60 min.	2 to 4 f Gm.
" Caryophyllus	5 to 10 min.	3 to 6 f D.
" Cascara Sagrada	15 to 60 min.	1 to 4 f Gm.
" Cascarilla	1 to 2 fl. dr.	4 to 8 f Gm.
" Cassia Fistula	1 to 4 fl. dr.	5 to 15 f Gm.
" " Mariland.	1 to 4 fl. dr.	5 to 15 f Gm.
" Castanea	1 to 2 fl. dr.	4 to 8 f Gm.
" Cataria	30 to 60 min.	2 to 4 f Gm.
" Catechu	8 to 30 min.	0.50 to 2 f Gm.
" Caulophyll.	25 to 40 min.	1 to 2.50 f Gm.
" Cereus Grandifl.	2 to 5 min.	1 to 4 f D.
" Chamælirium	15 to 40 min.	1 to 2.50 f Gm.
" Chelidonium	30 to 60 min.	2 to 4 f Gm.
" Chelone	1 to 2 fl. dr.	4 to 8 f Gm.
" Chenopod.	15 to 60 min.	1 to 4 f Gm.
" Chimaphila	1 to 2 fl. dr.	4 to 8 f Gm.
" Chionanthus	1 to 2 fl. dr.	4 to 8 f Gm.
" Chirata	30 to 60 min.	2 to 4 f Gm.

Remedies.	Dose in Apothecaries' Weights and Measures.	Dose in Metric Terms.
Fluidextr. Chondodendrum	½ to 2 fl. dr.	2 to 8 fGm.
" Cicuta	3 to 8 min.	2 to 6 fD.
" Cimicifuga	8 to 30 min.	0.50 to 20 fGm.
" Cinchona	30 to 60 min.	2 to 4 fGm.
" Coca	1 to 3 fl. dr.	4 to 12 fGm.
" Cocculus	1 to 3 min.	1 to 2 fD.
" Colch. Rad.	2 to 5 min.	1 to 4 fD.
" " Sem.	5 to 10 min.	3 to 6 fD.
" Collinsonia	30 to 60 min.	2 to 4 fGm.
" Colocynth.	5 to 10 min.	3 to 6 fD.
" Columbo	15 to 60 min.	1 to 4 fGm.
" Comptonia	30 to 60 min.	·2 to 4 fGm.
" Condurango	8 to 30 min.	0.50 to 2 fGm.
" Conium Fol.	3 to 10 min.	2 to 6 fD.
" " Fruct.	2 to 5 min.	1 to 4 fD.
" Convallaria	30 to 60 min.	2 to 4 fGm.
" Coptis	30 to 60 min.	2 to 4 fGm.
" Coto	1 to 5 min.	1 to 4 fD.
" Coriandrum	15 to 30 min.	1 to 2 fGm.
" Cornus Circin.	15 to 60 min.	1 to 4 fGm.
" " Flor.	30 to 60 min.	2 to 4 fGm.
" " Sericea	15 to 60 min.	1 to 4 fGm.
" Corydalis	15 to 60 min.	1 to 4 fGm.
" Cubeba	15 to 30 min.	1 to 2 fGm.
" Cusparia	15 to 40 min.	1 to 2.50 fGm.
" Cypriped.	15 to 60 min.	1 to 4 fGm.
" Damiana	1 to 2 fl. dr.	4 to 8 fGm.
" Delphinium	1 to 3 min.	1 to 2 fD.
" Digital. Fol.	2 to 5 min.	1 to 4 fD.
" " Rad.	1 to 3 min.	1 to 2 fD.
" Dioscorea	·15 to 40 min.	1 to 2.50 fD.
" Dipterix Odor.	5 to 10 min.	3 to 6 fD.
" Dita	1 to 4 fl. dr.	5 to 15 fGm.
" Dracont.	30 to 60 min.	2 to 4 fGm.
" Drosera	5 to 10 min.	3 to 6 fD.
" Dulcam.	1 to 2 fl. dr.	4 to 8 fGm.

Remedies.	Dose in Apothecaries' Weights and Measures.	Dose in Metric Terms.
Fluidextr. Epigæa	1 to 2 fl. dr.	4 to 8 f Gm.
" Ergota	15 to 60 min.	1 to 4 f Gm.
" Erigeron	1 to 2 fl. dr.	4 to 8 f Gm.
" Eriodictyon	15 to 30 min.	1 to 2 f Gm.
"˙ Erythroxylon	1 to 4 fl. dr.	5 to 15 f Gm.
" Eucalyptus	15 to 60 min.	1 to 4 f Gm.
" Euonymus	15 to 60 min.	1 to 4 f Gm.
" Eupatorium	30 to 60 min.	2 to 4 f Gm.
" Euphorb. Coroll.	3 to 10 min.	1 to 6 f D.
" " Ipecac.	5 to 15 min.	3 to 10 f D.
" Filix Mas	2 to 4 fl. dr.	6 to 15 f D.
" Fœniculum	15 to 30 min.	1 to 2 f Gm.
" Frangula	½ to 2 fl. dr.	2 to 8 f Gm.
" Frankenia	8 to 20 min.	5 to 12 f D.
" Frasera	30 to 60 min.	2 to 4 f Gm.
" Fuc. Vesicul.	½ to 2 fl. dr.	2 to 8 f Gm.
" Galanga	5 to 20 min.	3 to 12 f D.
" Galipea	15 to 40 min.	1 to 2.50 f Gm.
" Galium	1 to ·2 fl. dr.	4 to 8 f Gm.
" Galla	1 to 2 fl. dr.	4 to 8 f Gm.
" Gaültheria	1 to 2 fl. dr.	4 to 8 f Gm.
" Gelsem.	2 to 8 min.	1 to 5 f D.
" Gent.	30 to 60 min.	2 to 4 f Gm.
" " Comp.	30 to 60 min.	2 to 4 f Gm.
" " Catesbæi	30 to 60 min.	2 to 4 f Gm.
" Gent. Quinquefol.	8 to 40 min.	0.50 to 2.50 f Gm
" Geran.	15 to 60 min.	1 to 4 f Gm.
" Geum	15 to 60 min.	1 to 4 f Gm.
" Gillenia	8 to 30 min.	5 to 20 f D.
" Glycyrrh.	1 to 2 fl. dr.	4 to 8 f Gm.
" Gossypium	½ to 1½ fl. dr.	2 to 5 f Gm.
" Granat. Fruct. Cort.	1 to 2 fl. dr.	4 to 8 f Gm.
" " Rad. "	1 to 2 fl. dr.	4 to 8 f Gm.
" Grindel. Rob.	30 to 60 min.	2 to 4 f Gm.
" " Squarr.	1 to 2 fl. dr.	4 to 8 f Gm.
" Guaco	30 to 60 min.	2 to 4 f Gm.

Remedies.	Dose in Apothecaries' Weights and Measures.	Dose in Metric Terms.
Fluidextr. Guaiacum	30 to 60 min.	2 to 4 f Gm.
" Guarana	8 to 30 min.	5 to 20 f D.
" Hæmatox.	30 to 60 min.	2 to 4 f Gm.
" Hamamelis	1 to 2 fl. dr.	4 to 8 f Gm.
" Hedeoma	30 to 60 min.	2 to 4 f Gm.
" Helianthem.	1 to 2 fl. dr.	4 to 8 f Gm.
" Helleb. Nig.	8 to 20 min.	5 to 15 f D.
" Helonias	8 to 20 min.	5 to 15 f D.
" Hemidesmus	30 to 60 min.	2 to 4 f Gm.
" Hepatica	30 to 60 min.	2 to 4 f Gm.
" Heuchera	½ to 2 fl. dr.	2 to 8 f Gm.
" Humulus	30 to 60 min.	2 to 4 f Gm.
" Hydrangea	30 to 60 min.	2 to 4 f Gm.
" Hydrastis	8 to 30 min.	5 to 20 f D.
" Hyoscyam. Fol.	5 to 10 min.	3 to 6 f D.
" " Sem.	3 to 6 min.	2 to 4 f D.
" Hypericum	½ to 2 fl. dr.	2 to 8 f Gm.
" Hyssopus	30 to 60 min.	2 to 4 f Gm.
" Ignatia	1 to 5 min.	1 to 3 f D.
" Ilex	30 to 60 min.	2 to 4 f Gm.
" Inula	15 to 60 min.	1 to 4 f Gm.
" Ipecac.	3 to 60 min.	2 to 40 f D.
" Iris Flor.	8 to 60 min.	5 to 40 f D.
" " Versicol.	½ to 1 fl. dr.	2 to 4 f Gm.
" Jacaranda Proc.	15 to 60 min.	1 to 4 f Gm.
" Jaborandi	15 to 60 min.	1 to 4 f Gm.
" Jalapa	15 to 60 min.	1 to 4 f Gm.
" Juglans	1 to 2 fl. dr.	4 to 8 f Gm.
" Junip. Fruct.	30 to 60 min.	2 to 4 f Gm.
" " Virgin.	1 to 2 fl. dr.	4 to 8 f Gm.
" Kalmia	15 to 30 min.	1 to 2 f Gm.
" Kamala	1 to 2 fl. dr.	4 to 8 f Gm.
" Kava Kava	15 to 60 min.	1 to 4 f Gm.
" Kino	15 to 30 min.	1 to 2 f Gm.
" Kousso	2 to 4 fl. dr.	8 to 15 f Gm.
" Krameria	30 to 60 min.	2 to 4 f Gm.

Remedies.	Dose in Apothecaries' Weights and Measures.	Dose in Metric Terms.
Fluidextr. Lactuca	15 to 60 min.	1 to 4 f Gm.
" Lactucarium	8 to 20 min.	5 to 12 f D.
" Lappa	1 to 2 fl. dr.	4 to 8 f Gm.
" Larix	½ to 2 fl. dr.	2 to 8 f Gm.
" Leonurus	30 to 60 min.	2 to 4 f Gm.
" Leptandra	30 to 60 min.	2 to 4 f Gm.
" Liatris	30 to 60 min.	2 to 4 f Gm.
" Ligusticum	30 to 60 min.	2 to 4 f Gm.
" Liriodendron	½ to 2 fl. dr.	2 to 8 f Gm.
" Lobelia Herba	5 to 30 min.	3 to 20 f D.
" " Sem.	5 to 20 min.	3 to 15 f D.
" Lupulinum	5 to 10 min.	3 to 6 f D.
" Lycopus	1 to 4 fl. dr.	5 to 15 f Gm.
" Magnolia	30 to 60 min.	2 to 4 f Gm.
" Malva	2 to 8 fl. dr.	8 to 30 f Gm.
" Manzanita	·1 to 2 fl. dr.	4 to 8 f Gm.
" Marrubium	1 to 2 fl. dr.	4 to 8 f Gm.
" Matè	30 to 60 min.	2 to 4 f Gm.
" Matico	30 to 60 min.	2 to 4 f Gm
" Matricaria	8 to 30 min.	5 to 20 f D.
" Melissa	2 to 4 fl. dr.	8 to 15 f Gm.
" Menispermum	30 to 60 min.	2 to 4 f Gm.
" Mentha	1 to 2 fl. dr.	4 to 8 f Gm.
" Mezereum	5 to 15 min.	3 to 10 f D.
" Micromeria	15 to 60 min.	1 to 4 f Gm.
" Mikania	½ to 1 fl. dr.	2 to 4 f Gm.
" Mitchella	30 to 60 min.	2 to 4 f Gm.
" Monarda	5 to 20 min.	3 to 12 f D.
" Myrica	30 to 60 min.	2 to 4 f Gm.
" Nectandra	1 to 4 fl. dr.	5 to 15 f Gm.
" Nux Vom.	1 to 5 min.	1 to 4 f D.
" Nymphæa	15 to 60 min.	1 to 4 f Gm.
" Oenothera	15 to 30 min.	1 to 2 f Gm.
" Opium	8 to 40 min.	5 to 25 f D.
" Orobanche	8 to 30 min.	5 to 20 f D.
" Panax	½ to 2 fl. dr.	2 to 8 f Gm.

Remedies.	Dose in Apothecaries' Weights and Measures.	Dose in Metric Terms.
Fluidextr. Papaver	½ to 2 fl. dr.	2 to 8 f Gm.
" Pareira	30 to 60 min.	2 to 4 f Gm.
" Paullinia	8 to 30 min.	5 to 20 f D.
" Penthorum	8 to 20 min.	5 to 15 f D.
" Pepo	1 to 2 fl. dr.	4 to 8 f Gm.
" Petroselin.	½ to 3 fl. dr.	2 to 10 f Gm.
" Peumus Boldus	5 to 15 min.	3 to 10 f D.
" Phellandr.	2 to 10 min.	1 to 6 f D.
" Phoradendron Flav	30 to 60 min.	2 to 4 f Gm.
" Physostigma	1 to 3 min.	1 to 2 f D.
" Phytolacca Baccæ	10 to 30 min.	6 to 20 f D.
" " Rad.	10 to 30 min.	6 to 20 f D.
" Pilocarpus	15 to 60 min.	1 to 4 f Gm.
" Pimenta	10 to 40 min.	6 to 25 f D.
" Piper Cubeb.	15 to 30 min.	1 to 2 f Gm.
" " Nigr.	5 to 20 min.	3 to 15 f D.
" " Methyst.	15 to 60 min.	1 to 4 f Gm.
" Piscidia Erythrina	½ to 2 fl. dr.	2 to 8 f Gm.
" Podophyllum	8 to 20 min.	5 to 15 f D.
" Polygala Rubella	1 to 3 fl. dr.	4 to 10 f Gm.
" Polygon.	15 to 30 min.	1 to 2 f Gm.
" Polymnia	3 to 6 min.	2 to 4 f D.
" Polytrichum	1 to 2 fl. dr.	4 to 8 f Gm.
" Populus	30 to 60 min.	2 to 4 f Gm.
" Potentilla	30 to 60 min	2 to 4 f Gm.
" Prinos	30 to 60 min.	2 to 4 f Gm.
" Prun. Virg.	30 to 60 min.	2 to 4 f Gm.
" " " Co.	30 to 60 min	2 to 4 f Gm.
" Ptelea	8 to 30 min.	5 to 20 f D.
" Pterocaulon	15 to 60 min.	1 to 4 f Gm.
" Pulegium	1 to 4 fl. dr.	5 to 15 f Gm.
" Pulmonaria	30 to 60 min.	2 to 4 f Gm.
" Pulsatilla	2 to 5 min.	1 to 3 f D.
" Pycnanthemum	1 to 2 fl. dr.	4 to 8 f Gm.
" Pyrethrum	30 to 60 min.	2 to 4 f Gm.
" Quassia	30 to 60 min.	2 to 4 f Gm.

Remedies.	Dose in Apothecaries' Weights and Measures.	Dose in Metric Terms.
Fluidextr. Quercus	30 to 60 min.	2 to 4 f Gm.
" Rhamnus Cath.	30 to 60 min.	2 to 4 f Gm.
" " Frang.	1 to 2 fl. dr.	4 to 8 f Gm.
" " Pursh.	15 to 60 min.	1 to 4 f Gm.
" Rheum	15 to 40 min.	1 to 2.50 f Gm.
" Rhododendron	8 to 30 min.	5 to 20 f D.
" Rhus Glabra	30 to 60 min.	2 to 4 f Gm.
" " Toxicod.	1 to 6 min.	1 to 4 f D.
" Ricinus Folia	1 to 4 fl. dr.	5 to 15 f Gm.
" " Semina	5 to 10 min.	3 to 6 f D.
" Rosa Gall.	30 to 60 min.	2 to 4 f Gm.
" Rottlera	1 to 2 fl. dr.	4 to 8 f Gm.
" Rubia	30 to 60 min.	2 to 4 f Gm.
" Rubus Triv.	15 to 60 min.	1 to 4 f Gm.
" " Villos.	1 to 2 fl. dr.	4 to 8 f Gm.
" Rudbeckia	1 to 2 fl. dr.	4 to 8 f Gm.
" Rumex Crisp.	30 to 60 min.	2 to 4 f Gm.
" Ruta	8 to 30 min.	5 to 20 f D.
" Sabbatia	30 to 60 min.	2 to 4 f Gm.
" Sabina	5 to 20 min.	3 to 15 f D.
" Salix	½ to 2 fl. dr.	2 to 8 f Gm.
" Salvia	15 to 30 min.	1 to 2 f Gm.
. " Sambucus	1 to 2 fl. dr.	4 to 8 f Gm.
" Sanguin.	8 to 15 min.	5 to 10 f D.
" Santal. Alb.	1 to 2 fl. dr.	4 to 8 f Gm.
" Santonica	8 to 30 min.	5 to 20 f D.
" Sarracenia	15 to 30 min.	1 to 2 f Gm.
" ´ Sarsap.	1 to 2 fl. dr.	4 to 8 f Gm.
" " Co.	1 to 2 fl. dr.	4 to 8 f Gm.
" Sassafras	30 to 60 min.	2 to 4 f Gm.
" Scilla	5 to 60 min.	3 to 40 f D.
" Scoparius	30 to 60 min.	2 to 4 f Gm.
" Scutellaria	1 to 2 fl. dr.	4 to 8 f Gm.
" Senecio	1 to 2 fl. dr.	4 to 8 f Gm.
" Senega	8 to 20 min.	5 to 15 f D.
" Senna	1 to 4 fl. dr.	5 to 15 f Gm.

Remedies.	Dose in Apothecaries' Weights and Measures.	Dose in Metric Terms.
Fluidextr. Serpentar.	30 to 60 min.	2 to 4 f Gm.
" Silphium	15 to 40 min.	1 to 2.50 f Gm.
" Simaba Cedron	1 to 8 min.	1 to 5 f D.
" Simaruba	30 to 60 min.	2 to 4 f Gm.
" Solidago	30 to 60 min.	2 to 4 f Gm.
" Spigelia	30 to 60 min.	2 to 4 f Gm.
" Spigel. Senna	1 to 4 fl. dr.	5 to 15 f Gm.
" Spiræa	30 to 60 min.	2 to 4 f Gm.
" Statice	30 to 60 min.	2 to 4 f Gm.
" Stillingia	1 to 2 fl. dr.	4 to 8 f Gm.
" " Co.	1 to 2 fl. dr.	4 to 8 f Gm.
" Stramon. Fol.	2 to 5 min.	1 to 3 f D.
" " Sem.	1 to 3 min.	1 to 2 f D.
" Sumbul	15 to 60 min.	1 to 4 f Gm.
" Symphyt.	30 to 60 min.	2 to 4 f Gm.
" Tanacet.	30 to 60 min.	2 to 4 f Gm.
" Taraxac.	1 to 2 fl. dr.	4 to 8 f Gm.
" " Senna	½ to 2 fl. dr.	2 to 8 f Gm.
" Taxus Bacc.	1 to 5 min.	1 to 3 f D.
" Thea	1 to 3 fl. dr.	5 to 12 f Gm.
" Thuja	8 to 30 min.	5 to 20 f D.
" Thymus	30 to 60 min.	2 to 4 f Gm.
" Tilia	1 to 3 fl. dr.	4 to 12 f Gm.
" Tormentilla	15 to 30 min.	1 to 2 f Gm.
" Toxicodendron	1 to 5 min.	1 to 3 f D.
" Trifol. Prat.	1 to 2 fl. dr.	4 to 8 f Gm.
" Trillium	1 to 2 fl. dr.	4 to 8 f Gm.
" Triosteum	15 to 30 min.	1 to 2 f Gm.
" Tritic. Rep.	2 to 4 fl. dr.	8 to 15 f Gm.
" Tussilago	30 to 60 min.	2 to 4 f Gm.
" Urtica	5 to 20 min.	3 to 15 f D.
" Ustilago Maidis	1 to 20 min.	1 to 15 f D.
" Uva Ursi	30 to 60 min.	2 to 4 f Gm.
" Vaccin. Crassifol.	30 to 60 min.	2 to 4 f Gm.
" Valer.	15 to 30 min.	1 to 2 f Gm.
" Vanilla	3 to 10 min.	2 to 6 f D.

Remedies.	Dose in Apothecaries' Weights and Measures.	Dose in Metric Terms.
Fluidextr. Veratr. Alb.	2 to 5 min.	1 to 3 f D.
"　　Veratr. Vir.	2 to 5 min.	1 to 3 f D.
"　　Verbasc.	1 to 4 fl. dr.	5 to 15 f Gm.
"　　Verbena	15 to 60 min.	1 to 4 f Gm.
"　　Viburn Opul.	1 to 3 fl. dr.	4 to 12 f Gm.
"　　"　Prunif.	1 to 2 fl. dr.	4 to 8 f Gm.
"　　Viola Pedata	30 to 60 min.	2 to 4 f Gm.
"　　"　Tricol.	1 to 4 fl. dr.	5 to 15 f Gm.
"　　Viscum Alb.	30 to 60 min.	2 to 4 f Gm.
"　　Xanthium	8 to 20 min.	5 to 12 f D.
"　　Xanthorrhiza	30 to 60 min.	2 to 4 f Gm.
"　　Xanthoxylum	8 to 30 min.	5 to 20 f D.
"　　Yerba Reuma	8 to 20 min.	5 to 15 f D.
"　　"　Santa	15 to 60 min.	1 to 4 f Gm.
"　　Zedoaria	15 to 60 min.	1 to 4 f Gm.
"　　Zingiber	5 to 40 min.	3 to 30 f D.
Fœniculum	15 to 36 gr.	1 to 2 Gm.
Frangula Decoct.	2 to 8 fl. dr.	8 to 30 f Gm.
"　Extr.	5 to 15 gr.	0.30 to 1 Gm.
Frankenia ·	10 to 20 gr.	60 to 120 cents.
Frasera	30 to 60 gr.	2 to 4 Gm.
Fraserin (eclectic)	1 to 3 gr.	5 to 20 cents.
Fucus Vesicul.	15 to 60 gr.	1 to 4 Gm.
Galanga	5 to 15 gr.	30 to 100 cents.
Galbanum	5 to 20 gr.	30 to 120 cents.
Galium	30 to 60 gr.	2 to 4 Gm.
Galla	5 to 20 gr.	30 to 120 cents.
"　Tinct.	½ to 2 fl dr.	2 to 8 f Gm.
Gambogium	1 to 3 gr.	5 to 20 cents.
Gaultheria	½ to 2 dr.	2 to 8 Gm.
Gelsemin. (eclectic)	½ to 1 gr.	3 to 6 cents.
Gelsemium Extr.	2 to 5 gr.	10 to 30 cents.
"　Tinct.	5 to 15 min.	3 to 10 f D.
Gentiana Extr.	5 to 20 gr.	30 to 120 cents.
"　Inf.	1 to 2 fl. oz.	30 to 60 f Gm.
"　Tinct. Co.	½ to 2 fl. dr.	2 to 8 f Gm.

Remedies.	Dose in Apothecaries' Weights and Measures.	Dose in Metric Terms.
Gentiana Catesbæi	30 to 60 gr.	2 to 4 Gm.
" Quinquefol.	30 to 60 gr.	2 to 4 Gm.
Geraniin (eclectic)	1 to 3 gr.	5 to 20 cents.
Geranium	15 to 60 gr.	1 to 4 Gm.
Geum	15 to 60 gr.	1 to 4 Gm.
Gillenia	15 to 30 gr.	1 to 2 Gm.
Glycerin	½ to 2 fl. dr.	2 to 8 f Gm.
Glycer. Acid. Carbol.	5 to 10 min.	3 to 6 f D.
" " Gall.	15 to 60 min.	1 to 4 f Gm.
" " Tann.	8 to 40 min.	5 to 25 f D.
" Pix Liqu.	½ to 2 fl. dr.	2 to 8 f Gm.
Glycyrrh.	1 to 3 dr.	4 to 12 Gm.
" Extr.	5 to 30 gr.	30 cts. to 2 Gm.
" Mist. Co.	1 to 4 fl. dr.	5 to 15 f Gm.
" Pulv. Co.	8 to 60 gr.	5 to 40 f D.
Glycyrrhizinum	5 to 10 gr.	30 to 60 cents.
Gossypiin (eclectic)	1 to 5 gr.	5 to 30 cents.
Gossypium Rad.	30 to 60 gr.	2 to 4 Gm.
Granat. Fruct.	15 to 30 gr.	1 to 2 Gm.
" Rad. Cort.	8 to 30 gr.	50 cts to 2 Gm.
" " " Dec.	1 to 2 fl. oz.	30 to 60 f Gm.
Grindel. Rob.	30 to 60 gr.	2 to 4 Gm.
" " Tinct.	2 to 8 fl. dr.	8 to 30 f Gm.
" Squarr.	30 to 60 gr.	2 to 4 Gm.
Guaco	30 to 60 gr.	2 to 4 Gm.
Guaiacum Lign. Dec.	1 to 2 fl. oz.	30 to 60 f Gm.
" Resina	5 to 30 gr.	30 cts. to 2 Gm.
" " Tinct.	30 to 60 min.	2 to 4 f Gm.
" " " Co.	30 to 60 min.	2 to 4 f Gm.
Guarana	8 to 30 gr.	50 cts. to 2 Gm.
" Tinct.	1 to 8 fl. dr.	5 to 30 f Gm.
Gynocard. Oleum	4 to 16 min.	3 to 12 f D.
Hæmatoxylum	½ to 2 dr.	2 to 8 Gm.
" Dec.	1 to 2 fl. oz.	30 to 60 f Gm.
" Extr.	8 to 30 gr.	50 to 200 cents
Hamamelis	½ to 2 dr.	2 to 8 Gm.

Remedies.	Dose in Apothecaries' Weights and Measures.	Dose in Metric Terms.
Hamamelin (eclectic)	1 to 3 gr.	5 to 20 cents.
Hedeoma	1 to 4 dr.	5 to 15 Gm.
Helianthem.	5 to 20 gr.	30 to 120 cents.
Helleb. Nig.	5 to 20 gr.	30 to 120 cents.
" " Tinct.	30 to 60 min.	2 to 4 Gm.
Helonias	15 to 40 gr.	1 to 2.50 fGm.
Helonin (eclectic)	2 to 4 gr.	10 to 25 cents.
Hemidesmus	30 to 60 gr.	2 to 4 Gm.
Hepatica	30 to 60 gr.	2 to 4 Gm.
Heuchera	$\frac{1}{2}$ to 2 dr.	2 to 8 Gm.
Hordeum Dec.	1 to 8 fl. oz.	30 to 250 fGm.
Humulus	15 to 60 gr.	1 to 4 Gm.
" Extr.	5 to 20 gr.	30 to 120 cents.
" Inf.	1 to 2 fl. oz.	30 to 60 fGm.
" Tinct.	1 to 3 fl. dr.	30 to 100 fGm.
Hydrangea	30 to 60 gr.	2 to 4 Gm.
Hydrarg. Chlorid. Corros.	$\frac{1}{32}$ to $\frac{1}{6}$ gr.	2 to 10 mills.
" " Mite	$\frac{1}{6}$ to 8 gr.	1 to 50 cents.
" Cyanid.	$\frac{1}{16}$ to $\frac{1}{2}$ gr.	4 to 30 mills.
" Iodid. Flav.	$\frac{1}{2}$ to 3 gr.	3 to 20 cents.
" " Rubr.	$\frac{1}{16}$ to $\frac{1}{4}$ gr.	4 to 15 mills.
" " Vir.	$\frac{1}{2}$ to 3 gr.	3 to 20 cents.
" Oxid. Flav.	$\frac{1}{12}$ to $\frac{1}{4}$ gr.	5 to 15 mills.
" " Rubr.	$\frac{1}{16}$ to $\frac{1}{2}$ gr.	4 to 30 mills.
" " Nigr.	$\frac{1}{2}$ to 3 gr.	3 to 20 cents.
" Sulph. Flav.	$\frac{1}{4}$ to 1 gr.	15 to 60 mills.
" c. Creta	3 to 8 gr.	15 to 50 cents.
" Massa	3 to 15 gr.	15 cts. to 1 Gm.
Hydrastin (eclectic)	3 to 5 gr.	15 to 30 cents.
Hydrastis	$\frac{1}{2}$ to 2 dr.	2 to 8 Gm.
" Extr.	5 to 15 gr.	30 to 100 cents.
" Tinct.	1 to 8 fl. dr.	8 to 30 fGm.
Hydrogen. Peroxid. Sol.	1 to 4 fl. dr.	5 to 15 fGm.
Hyoscyamus Folia	5 to 10 gr.	30 to 60 cents.
" Fol. Extr.	2 to 3 gr.	10 to 20 cents.
" " " Alcoh.	1 to 2 gr.	5 to 12 cents.

Remedies.	Dose in Apothecaries' Weights and Measures.	Dose in Metric Terms.
Hyoscyamus Fol. Tinct.	15 to 60 min.	1 to 4 f Gm.
" Sem.	3 to 8 gr.	15 to 50 cents.
" " Extr.	1 to 3 gr.	5 to 20 cents.
" " Tinct.	8 to 45 min.	5 to 25 f D.
Hyoscyamina (and salts)	$\frac{1}{60}$ to $\frac{1}{24}$ gr.	1 to 3 mills.
Hyoscyamin (eclectic)	$\frac{1}{12}$ to $\frac{1}{2}$ gr.	5 to 30 mills.
Hyperic Perfor.	30 to 120 gr.	2 to 8 Gm.
Hyssopus	1 to 2 dr.	4 to 8 Gm.
Ignatia	1 to 2 gr.	5 to 20 cents.
" Extr.	$\frac{1}{2}$ to $1\frac{1}{2}$ gr.	3 to 25 cents.
" Tinct.	8 to 30 min.	5 to 20 f D.
Ilex	30 to 60 gr.	2 to 4 Gm.
Illicium	8 to 30 gr.	50 cts. to 2 Gm.
Infusum Angust.	1 to 2 fl. oz.	30 to 60 f Gm.
" Anthemis	1 to 3 "	30 to 100 f Gm.
" Aurant.	1 to 2 "	30 to 60 "
" " Comp.	1 to 2 "	30 to 60 "
" Barosma	1 to 2 "	30 to 60 "
" Brayera	4 to 8 "	100 to 250 "
" Buchu	1 to 2 "	30 to 60 "
" Calumba	1 to 2 "	30 to 60 "
" Capsicum	1 to 2 "	30 to 60 "
" Caryoph.	1 to 2 "	30 to 60 "
" Cascarilla	1 to 2 "	30 to 60 "
" Catechu	1 to 2 "	30 to 60 "
" " Co.	1 to 2 "	30 to 60 "
" Chirata	1 to 2 "	30 to 60 "
" Cinch. Fl.	1 to 2 "	30 to 60 "
" " Rub.	1 to 2 "	30 to 60 "
" Coptis	1 to 2 "	30 to 60 "
" Cusparia	1 to 2 "	30 to 60 "
" Digital.	2 to 4 fl. dr.	10 to 15 "
" Dulcam.	1 to 2 fl. oz.	30 to 60 "
" Ergota	1 to 2 "	30 to 60 "
" Eupator.	1 to 2 "	30 to 60 "
" Frasera	1 to 2 "	30 to 60 "

Remedies.	Dose in Apothecaries' Weights and Measures.	Dose in Metric Terms.
Inf. Gent. Co.	1 to 2 fl. oz.	30 to 60 f Gm.
" Humuli	1 to 2 "	30 to 60 "
" Ipecac.	8 to 60 min.	0.50 to 4 "
" Junip.	2 to 3 fl. oz.	60 to 100 "
" Kousso	4 to 8 "	100 to 250 "
" Kramer.	1 to 2 "	30 to 60 "
" Linum Co.	2 to 8 "	60 to 250 "
" Lupulinum	1 to 2 "	30 to 60 "
" Matico	1 to 2 "	30 to 60 "
" Pareira	1 to 2 "	30 to 60 "
" Pix Liqu.	2 to 4 "	60 to 120 "
" Prun. Virg.	2 to 3 "	60 to 100 "
" Quassia	1 to 2 "	30 to 60 "
" Rheum	1 to 2 "	30 to 60 "
" Rosa Co.	1 to 2 "	30 to 60 "
" Sabbatia	1 to 2 "	30 to 60 "
" Salvia	1 to 2 "	30 to 60 "
" Senega	1 to 2 "	30 to 60 "
" Senna	1 to 2 "	30 to 60 "
" " Co.	1 to 2 "	30 to 60 "
" Serpent.	1 to 2 "	30 to 60 "
" Spigel.	4 to 8 "	100 to 250 "
" Tabaci	4 to 8 "	100 to 250 "
" Taraxac.	1 to 2 "	30 to 60 "
" Uva Ursi	1 to 2 "	30 to 60 "
" Valer.	1 to 2 "	30 to 60 "
" Zingib.	1 to 2 "	30 to 60 "
Inula	15 to 60 gr.	1 to 4 Gm.
Iodinium	$\frac{1}{4}$ to $\frac{1}{2}$ gr.	15 to 30 mills.
" Tinct.	1 to 10 min.	1 to 6 f D.
" " Co.	5 to 15 min.	3 to 10 f D.
Iodoformum	1 to 3 gr.	5 to 20 cents.
Ipecacuanha { expect.	$\frac{1}{3}$ to 2 gr.	3 to 10 cents.
{ emet.	15 to 30 gr.	1 to 2 Gm.
" Inf.	8 to 60 min.	5 to 20 f D.
" Pulv. Co.	5 to 15 gr.	30 to 100 cents.

Remedies.		Dose in Apothecaries' Weights and Measures.	Dose in Metric Terms.
Ipecacuanha Syr.	{ expect. { emet.	30 to 60 min. 4 to 8 fl. dr.	2 to 4 f Gm. 15 to 30 f Gm.
" Tinct.	{ expect. { emet.	5 to 30 min. 2 to 6 fl. dr.	3 to 10 f D. 8 to 25 f Gm.
" Vin.	{ expect. { emet.	5 to 30 min. 2 to 6 fl. dr.	3 to 10 f D. 8 to 25 f Gm.
Iris Flor.		1 to 6 dr.	4 to 25 Gm.
" Versicol.		5 to 15 gr.	30 cts. to 1 Gm.
Irisin (eclectic)		2 to 4 gr.	10 to 25 cents.
Jaborandi		5 to 60 gr.	30 cts. to 4 Gm.
" Tinct.		½ to 2 fl. dr.	2 to 8 f Gm.
Jalapa		15 to 30 gr.	1 to 2 Gm.
" Extr. Aquos.		8 to 15 gr.	50 cts. to 1 Gm.
" " Alcoh.		3 to 5 gr.	15 to 30 cents.
" Pulv. Co.		8 to 30 gr.	50 cts. to 2 Gm.
" Resina		2 to 5 gr.	10 to 30 cents.
" Tinct.		½ to 2 fl. dr.	2 to 8 f Gm.
Jalapin (eclectic)		1 to 3 gr.	5 to 20 cents.
Juglandin (eclectic)		2 to 5 gr.	10 to 30 cents.
Juglans		30 to 60 gr.	2 to 4 Gm.
Junip. Fruct.		1 to 2 dr.	4 to 8 Gm.
" Virgin.		1 to 2 dr.	4 to 8 Gm.
Kalmia		15 to 30 gr.	1 to 2 Gm.
Kamala		1 to 2 dr.	4 to 8 Gm.
Kava Kava		30 to 60 gr.	2 to 4 Gm.
Kermes Miner.		1 to 3 gr.	5 to 20 cents.
Kino.		8 to 30 gr.	50 cts. to 2 Gm.
" Tinct.		½ to 2 fl. dr.	2 to 8 f Gm.
Koussin		8 to 30 gr.	50 cts. to 2 Gm.
Kousso Inf.		4 to 8 fl. oz.	100 to 250 f Gm.
Krameria		8 to 30 gr.	50 cts. to 2 Gm.
" Extr.		5 to 20 gr.	30 to 120 cents.
" Inf.		1 to 2 fl. oz.	30 to 60 f Gm.
" Syrup.		½ to 4 fl. dr.	2 to 15 f Gm.
" Tinct.		½ to 2 fl. dr.	2 to 8 f Gm.
Lactuca Extr.		1 to 5 gr.	5 to 30 cents.

Remedies.	Dose in Apothecaries' Weights and Measures.	Dose in Metric Terms.
Lactucarium	8 to 20 gr.	50 to 120 cents.
" Syrup.	2 to 3 fl. dr.	8 to 12 f Gm.
Lappa	30 to 60 gr.	2 to 4 Gm.
Larix	½ to 2 dr.	2 to 8 Gm.
Leontodin (eclectic)	2 to 4 gr.	10 to 25 cents.
Leonurus	1 to 2 dr.	4 to 8 Gm.
Leptandra	15 to 60 gr.	1 to 4 Gm.
" Extr.	3 to 8 gr.	15 to 50 cents.
" Resina	2 to 4 gr.	10 to 25 cents.
" Tinct.	2 to 6 fl. dr.	8 to 25 f Gm.
Leptandrin (eclectic)	2 to 4 gr.	10 to 25 cents.
Liatris	30 to 60 gr.	2 to 4 Gm.
Ligusticum	30 to 60 gr.	2 to 4 Gm.
Limon. Succus	½ to 4 fl. oz.	15 to 120 f Gm.
Linum Infus.	2 to 8 fl. oz.	60 to 250 f Gm.
Liquor Ammon. Acet.	2 to 8 fl. dr.	60 to 250 f Gm.
" " Anis.	8 to 30 min.	5 to 20 f D.
" Arsen. Chlor.	2 to 8 min.	1 to 5 f D.
" Arsen. et Hydrarg. Iod.	2 to 10 min.	1 to 6 f D.
" Bar. Chlor.	1 to 5 min.	1 to 4 f D.
" Bismuth. Ammon. Citr.	30 to 60 min.	2 to 4 f Gm.
" Calx Chlorid.	30 to 60 min.	2 to 4 f Gm.
" Calx	2 to 4 fl. oz.	60 to 120 f Gm.
" " Chlorin.	15 to 60 min.	1 to 4 f Gm.
" " Sacch.	15 to 30 min.	1 to 2 f Gm.
" Chlorof. Co.	5 to 10 min.	3 to 6 f D.
" Ferr. Chlorid.	2 to 10 min.	1 to 6 f D.
" " Citr.	5 to 10 min.	3 to 6 f D.
" " Dialys.	5 to 30 min.	3 to 20 f D.
" " Nitr.	8 to 20 min.	5 to 15 f D.
" " Subsulph.	5 to 15 min.	3 to 10 f D.
" Iodum Co.	2 to 6 min	1 to 4 f D.
" Magnes. Citr. { lax.	4 to 6 fl. oz.	100 to 200 f Gm.
{ purg.	6 to 12 fl. oz.	105 to 400 f Gm.
" Morph. Acet.	8 to 30 min.	5 to 20 f D.
" " Citr.	5 to 15 min.	3 to 10 f D.

Remedies.	Dose in Apothecaries' Weights and Measures.	Dose in Metric Terms.
Liquor Morph. Mecon.	8 to 30 min.	5 to 20 f D.
" " Sulph., U. S. P.	1 to 2 fl. dr.	4 to 8 f Gm.
" " " Magendie's	5 to 15 min.	3 to 10 f D.
" Opii Co.	8 to 30 min.	5 to 20 f D.
" Pepsinum	2 to 4 fl. dr.	8 to 15 f Gm.
" Potassa	5 to 20 min.	3 to 15 f D.
" Potass. Arsen.	3 to 10 min.	2 to 6 f D.
" " Citr.	2 to 4 fl. dr.	8 to 15 f Gm.
" " Silic.	1 to 3 gr.	5 to 20 cents.
" Soda	5 to 20 min.	3 to 12 f D.
" " Chlor.	30 to 60 min.	2 to 4 f Gm.
" " Arsen.	3 to 10 min.	2 to 6 f D.
" " Silic.	1 to 3 gr.	5 to 20 cents.
Liriodendrum	½ to 2 dr.	5 to 8 Gm.
Lith. Benz.	2 to 5 gr.	10 to 30 cents.
" Bromid.	1 to 3 gr.	5 to 20 cents.
" Carb.	2 to 6 gr.	10 to 35 cents.
" " Chlor.	2 to 6 gr.	10 to 35 cents.
" " Citr.	2 to 10 gr.	10 to 60 cents.
" Salicyl.	2 to 8 gr.	10 to 50 cents.
Lobelia Herba	5 to 20 gr.	30 to 120 cents.
" " Extr.	1 to, 3 gr.	5 to 20 cents.
" " Inf.	1 to 4 fl. dr.	5 to 15 f Gm.
" " Tinct.	30 to 60 min.	2 to 4 f Gm.
" Sem.	5 to 15 gr.	30 cts. to 1 Gm.
" " Tinct.	30 to 60 min.	2 to 4 f Gm.
Lobelin (eclectic)	¼ to 3 gr.	2 to 20 cents.
Lobelina (alkaloid)	1/32 to ⅛ gr.	2 to 8 mills.
Lupulinum	5 to 10 gr.	30 to 60 cents.
" Oleoresina	2 to 5 gr.	10 to 30 cents.
" Tinct.	½ to 2 fl. dr.	2 to 8 f Gm.
Lycopodium	8 to 30 gr.	50 cts. to 2 Gm.
Lycopin (eclectic)	1 to 4 gr.	5 to 25 cents.
Lycopus	1 to 4 dr.	5 to 15 Gm.
Macis	5 to 20 gr.	30 cts.to 1.20Gm
Macrotin (eclectic)	1 to 3 gr.	5 to 20 cents.

Remedies.	Dose in Apothecaries' Weights and Measures.	Dose in Metric Terms.
Magnesia	15 to 60 gr.	1 to 4 Gm.
" Carb.	½ to 2 dr.	2 to 8 Gm.
" Citr.	2 to 8 dr.	8 to 30 Gm.
" Hypophosph.	5 to 20 gr.	30 to 120 cents.
" Sulphas	2 to 8 dr.	8 to 30 Gm.
" Sulphis	8 to 40 gr.	50 cts.to 2.50Gm
Magnolia	30 to 60 gr.	2 to 4 Gm.
Malva	2 to 8 dr.	8 to 30 Gm.
Maltum Extr.	2 to 4 dr.	8 to 15 Gm.
" Inf.	2 to 8 fl. oz.	8 to 30 f Gm.
Mangan. Carb.	2 to 10 gr.	10 to 60 cents.
" Citr.	2 to 8 gr.	10 to 50 cents.
" Hypophosph.	2 to 8 gr.	10 to 50 cents.
" Iodid.	2 to 8 gr.	10 to 50 cents.
" Oxid Nigr.	2 to 10 gr.	10 to 60 cents.
" Sulphas	2 to 10 gr.	10 to 60 cents.
Manna	1 to 2 oz.	30 to 60 Gm.
" Syrup.	2 to 8 oz.	60 to 250 Gm.
Manzanita	1 to 2 dr.	4 to 8 Gm.
Marrub.	1 to 2 dr.	4 to 8 Gm.
Mastiches	2 to 5 gr.	10 to 30 cents.
Maté	30 to 60 gr.	2 to 4 Gm.
Matico	15 to 60 gr.	1 to 4 Gm.
" Inf.	1 to 2 fl. oz.	30 to 60 f Gm.
" Tinct.	1 to 4 fl. dr.	5 to 15 f Gm.
Matricaria	8 to 30 gr.	50 cts. to 2 Gm.
" Inf.	2 to 8 fl. oz.	60 to 250 f Gm.
Melissa	2 to 4 dr.	8 to 15 Gm.
Menispermin (eclectic)	1 to 3 gr.	5 to 20 cents.
Menispermum	30 to 60 gr.	2 to 4 Gm.
Mentha Pip.	5 to 20 gr.	30 to 120 cents.
" " Aetherol.	1 to 4 min.	1 to 3 f D.
" " Aqua	1 to 2 fl. oz.	30 to 60 f Gm.
" " Spir.	30 to 60 min.	2 to 4 f Gm.
Mentha Virid.	5 to 20 gr.	30 to 120 cents.
Methyl. Iodid.	1 to 5 min.	1 to 3 f D.

Remedies.	Dose in Apothecaries' Weights and Measures.	Dose in Metric Terms.
Mezereum	5 to 10 gr.	30 to 60 cents.
Micromeria	30 to 60 gr.	2 to .4 Gm.
Mikania	30 to 60 gr.	2 to 4 Gm.
Mist. Ammon.	4 to 8 fl. dr.	100 to 250 f Gm.
" Amygd.	2 to 8 fl. oz.	60 to 250 f Gm.
" Asafœt.	4 to 8 fl. dr.	15 to 30 f Gm.
" Chlorof.	4 to 8 fl. dr.	15 to 30 f Gm.
" Copaiba	1 to 4 fl. dr.	4 to 15 f Gm.
" Creosotum	1 to 2 fl. oz.	30 to 60 f Gm.
" Creta	1 to 2 fl. oz.	30 to 60 f Gm.
" Ferr. Co.	1 to 2 fl. oz.	30 to 60 f Gm.
" Glycyrrh. Co.	1 to 4 fl. dr.	4 to 15 f Gm.
" Guaiacum	$\frac{1}{2}$ to 2 fl. oz.	15 to 60 f Gm.
" Myrrha Ferr.	1 to 2 fl. oz.	30 to 60 f Gm.
" Pot. Citr.	2 to 4 fl. oz.	60 to 120 f Gm.
" Rheum Co.(for children)	30 to 60 min.	2 to 4 f Gm.
" Scammon.	$\frac{1}{2}$ to 2 fl. oz.	15 to 60 f Gm.
" Senna Co.	1 to 2 fl. oz.	30 to 60 f Gm.
Mitchella	30 to 60 gr.	2 to 4 f Gm.
Monarda	5 to 20 gr.	30 to 120 cents.
Morphina	$\frac{1}{16}$ to $\frac{1}{2}$ gr.	4 to 30 mills.
Morph. Acet.	$\frac{1}{16}$ to $\frac{1}{2}$ gr.	4 to 30 mills.
" Hydrobrom.	$\frac{1}{16}$ to $\frac{1}{2}$ gr.	4 to 30 mills.
" Hydrochlor.	$\frac{1}{16}$ to $\frac{1}{2}$ gr.	4 to 30 mills.
" Sulph.	$\frac{1}{16}$ to $\frac{1}{2}$ gr.	4 to 30 mills.
" " Pulv. Co.	8 to 15 gr.	50 cts. to 1 Gm.
" Valer.	$\frac{1}{8}$ to $\frac{1}{2}$ gr.	8 to 30 mills.
Morrhua Oleum	1 to 8 fl. dr.	30 to 250 f Gm.
Moschus	$\frac{1}{2}$ to 4 gr.	3 to 25 cents.
" Tinct.	8 to 30 min.	5 to 20 f D.
Mucuna	15 to 20 gr.	1 to 1.20 Gm.
Myrica	30 to 60 gr.	2 to 4 Gm.
Myricin (eclectic)	1 to 3 gr.	5 to 20 cents.
Myristica	5 to 20 gr.	30 to 120 cents.
" Aetherol.	1 to 3 min.	5 to 20 cents.
Myrrha	8 to 30 gr.	50 cts. to 2 Gm.

Remedies.	Dose in Apothecaries' Weights and Measures.	Dose in Metric Terms.
Myrrha Tinct.	30 to 60 min.	2 to 4 f Gm.
" " Arom.	30 to 60 min.	2 to 4 f Gm.
Naphtalinum	½ to 3 gr.	2 to 20 cents.
Narceina	¼ to 1½ gr.	15 to 100 mills.
Narcotina	½ to 5 gr.	3 to 30 cents.
Nectandra	1 to 4 dr.	5 to 15 Gm.
Nicotina	$\frac{1}{120}$ to $\frac{1}{30}$ gr.	0.50 to 2 mills.
Nitroglycerin	$\frac{1}{64}$ to $\frac{1}{16}$ gr.	1 to 4 mills.
Nux Vomica	1 to 5 gr.	5 to 30 cents.
" Extr.	$\frac{1}{6}$ to 1 gr.	1 to 6 cents.
" Tinct.	8 to 30 min.	5 to 20 f D.
Nymphæa	15 to 45 gr.	1 to 3 Gm.
Oenothera	30 to 60 gr.	2 to 4 Gm.
Olea Aetherea or Volatilia *		
Oleoresina Capsic.	$\frac{1}{8}$ to 1 gr.	8 mills to 6 cts.
" Cubeb.	5 to 30 min.	3 to 20 f D.
" Filix	15 to 40 min.	10 to 25 f D.
" Lupul.	2 to 5 gr.	10 to 30 cents.
" Piper	1 to 3 min.	1 to 2 f D.
" Xanthoxyl.	2 to 5 min.	1 to 3 f D.
" Zingiber	$\frac{1}{8}$ to 1 gr.	8 mills to 6 cts.
Oleum Amygd. Expr.	1 to 8 fl. dr.	4 to 30 f Gm.
" Ergota "	1 to 5 min.	1 to 3 f D.
" Gossypium Sem.	1 to 8 fl. dr.	4 to 30 f Gm.
" Gynocardia	5 to 20 gr.	30 to 120 cents.
" Jecoris Aselli	1 to 8 fl. dr.	4 to 30 f Gm.
" Lini	½ to 2 fl. oz.	15 to 60 f Gm.
" Morrhua	1 to 8 fl. dr.	4 to 30 f Gm.
" Oliva	½ to 2 fl. oz.	15 to 60 f Gm.
" Phosphoratum	1 to 3 gr.	5 to 20 cents.
" Ricinus	1 to 8 fl. dr.	4 to 30 f Gm.
" Sesamum	½ to 2 fl. oz.	15 to 60 f Gm.
" Tiglium	$\frac{1}{4}$ to 2 gr.	1 to 10 cents.
Opium	$\frac{1}{8}$ to 2 gr.	8 mills to 10 cts.
" Acetum	3 to 10 min.	2 to 6 f D.
" Denarcot.	$\frac{1}{6}$ to 1½ gr.	1 to 10 cents.

* Volatile Oils will be found under " Aetherolcum."

Remedies.	Dose in Apothecaries' Weights and Measures.	Dose in Metric Terms.
Opium Extr.	$\frac{1}{6}$ to $\frac{1}{2}$ gr.	1 to 3 cents.
" Liqu. Co. (Squibb)	8 to 30 min.	5 to 20 f D.
" Pil.	1 to 2 pills	1 to 2 pills.
" Tinct.	5 to 30 min.	3 to 20 f D.
" " Acet.	5 to 20 min.	3 to 12 f D.
" " Ammon.	$\frac{1}{2}$ to $1\frac{1}{2}$ fl. dr.	2 to 6 f Gm.
" " Camph.	10 to 60 min.	6 to 40 f D.
" " Deodor.	5 to 25 min.	3 to 15 f D.
" Vinum	3 to 15 min.	2 to 10 f D.
" " Croc.	2 to 8 min.	1 to 5 f D.
Oxymel Scilla	$\frac{1}{2}$ to 2 fl. dr.	2 to 8 f Gm.
Panax	$\frac{1}{2}$ to 2 dr.	2 to 8 Gm.
Papaver	15 to 60 gr.	1 to 4 Gm.
" Syr.	2 to 8 fl. dr.	8 to 30 f Gm.
Papaverina	1 to 4 gr.	5 to 15 f Gm.
Paracotoinum	$\frac{1}{4}$ to 4 gr.	15 mills to 25 cts.
Pareira	$\frac{1}{2}$ to 2 dr.	2 to 8 Gm.
Paullinia Sem.	8 to 30 gr.	50 cts. to 2 Gm.
Penthorum	8 to 30 gr.	50 cts. to 2 Gm.
Pepo (in emulsion)	1 to 2 oz.	30 to 60 Gm.
Pepsinum	$\frac{1}{2}$ to 5 gr.	3 to 30 cents.
" Sacch.	2 to 15 gr.	10 cts. to 1 Gm.
" Liquor	2 to 4 fl. dr.	8 to 15 f Gm.
Perubalsam Syr.	1 to 3 fl. dr.	4 to 12 f Gm.
Petroselinum	$\frac{1}{2}$ to 3 dr.	2 to 12 Gm.
Peumus Boldus	8 to 15 gr.	50 to 100 cents.
Phellandr.	3 to 15 gr.	15 to 100 cents.
Phenol.	$\frac{1}{2}$ to 3 gr	3 to 20 cents.
Phlorizin	1 to 15 gr.	6 to 100 cents.
Phosphorus	$\frac{1}{100}$ to $\frac{1}{32}$ gr.	0.50 to 2 mills.
" Tinct.	8 to 40 min.	5 to 25 f D.
Phosphoratum Oleum	1 to 3 gr.	5 to 20 cents.
Physostigma	1 to 3 gr.	5 to 20 cents.
" Extr.	$\frac{1}{16}$ to $\frac{1}{4}$ gr.	4 to 15 mills.
" Tinct.	8 to 30 min.	5 to 20 f D.
Physostigmina (and salts)	$\frac{1}{120}$ to $\frac{1}{60}$ gr.	0.50 to 1 mill.

Remedies.	Dose in Apothecaries' Weights and Measures.	Dose in Metric Terms.
Phytolacca Bacca	8 to 30 gr.	50 cts. to 2 Gm.
" Rad.	8 to 30 gr.	50 cts. to 2 Gm.
" " Tinct.	1 to 4 fl. dr.	5 to 15 f Gm.
Phytolaccin (eclectic)	1 to 3 gr.	5 to 20 cents.
Picrotoxin	$\frac{1}{64}$ to $\frac{1}{8}$ gr.	1 to 8 mills.
Pilocarpina (and salts)	$\frac{1}{8}$ to $\frac{3}{4}$ gr.	8 to 50 mills.
Pilocarpus	8 to 60 gr.	50 cts. to 4 Gm.
" Tinct.	1 to 6 fl. dr.	5 to 25 f Gm.
Pilulæ Aloe	1 to 3 pills	1 to 3 pills.
" " Asafœt.	2 to 5 pills	2 to 5 pills.
" " Ferr.	5 to 10 gr.	30 to 60 cents.
" " Mastich.	1 to 2 pills	1 to 2 pills.
" " Myrrha	3 to 6 pills	3 to 6 pills.
" Antimon. Co.	1 to 3 pills	1 to 3 pills.
" Asafœt.	1 to 3 pills	1 to 3 pills.
" Cathart. Co.	1 to 4 pills	1 to 4 pills.
" " Imp.	1. to 5 pills	1 to 5 pills.
" Coloc. Co.	5 to 20 gr.	30 to 120 cents.
" Conium Co.	5 to 10 gr.	30 to 60 cents.
" Copaiba	2 to 6 pills	2 to 6 pills.
" Ferr. Brom.	3 to 10 gr.	15 to 60 cents.
" " Carb.	2 to 10 gr.	10 to 60 cents.
" " Co.	· 2 to 6 pills	2 to 6 pills.
" " Iodid.	1 to 5 pills	1 to 5 pills.
" Galb. Co.	1 to 5 pills	1 to 5 pills.
" Hydrarg.	3 to 15 gr.	15 to 100 cents.
" Opium	1 to 2 pills	1 to 2 pills.
" " Camph.	1 to 2 pills	1 to 2 pills.
" Phosphorus (Br. P.)	2 to 5 gr.	5 to 30 cents.
" Plumb. Acet. Opium	2 to 3 gr.	10 to 20 cents.
" Quin. Sulph.	1 to 5 pills	1 to 5 pills.
" Rheum	1 to 5 pills	1 to 5 pills.
" " Co.	2 to 4 pills	2 to 4 pills.
" Sapo Co.	1 to 10 gr.	5 to 60 cents.
" Scammon. Co.	5 to 25 gr.	30 to 100 cents.
" Scilla Co.	1 to 3 pills	1 to 3 pills.

Remedies.	Dose in Apothecaries' Weights and Measures.	Dose in Metric Terms.
Pimenta	8 to 30 gr.	50 cts. to 2 Gm.
Piper Nigr.	5 to 20 gr.	30 to 120 cents.
" Conf.	1 to 2 dr.	4 to 8 Gm.
Piperinum	1 to 8 gr.	5 to 50 cents.
Piper Methyst.	30 to 60 gr.	2 to 4 Gm.
Pix Liquida	8 to 60 gr.	50 cts. to 4 Gm.
Plat. Chlorid.	$\frac{1}{16}$ to $\frac{1}{2}$ gr.	4 to 15 mills.
Plumb. Acet.	$\frac{1}{2}$ to 3 gr.	15 to 50 mills.
" Iodid.	$\frac{1}{2}$ to 4 gr.	3 to 25 cents.
Podophyllum	5 to 30 gr.	30 cts. to 2 Gm.
" Extr.	5 to 15 gr.	30 cts. to 1 Gm.
" Resina	$\frac{1}{8}$ to $\frac{1}{2}$ gr.	8 to 30 mills.
Podophyllin (eclectic)	$\frac{1}{8}$ to $\frac{1}{2}$ gr.	8 to 30 mills.
Polygala	1 to 3 dr.	4 to 12 Gm.
Polygon. Punct.	15 to 30 gr.	1 to 2 Gm.
Polymnia	5 to 10 gr.	30 to 60 cents.
" Extr.	$\frac{1}{4}$ to $\frac{1}{2}$ gr.	1 to 3 cents.
Polytrich.	1 to 2 dr.	4 to 8 Gm.
Populus	$\frac{1}{2}$ to 2 dr.	2 to 8 Gm.
Populin (eclectic)	2 to 4 gr.	10 to 25 cents.
Potass. Acet.	1 to 3 dr.	4 to 12 Gm.
" Arsen.	$\frac{1}{32}$ to $\frac{1}{12}$ gr.	2 to 5 mills.
" " Liqu.	2 to 8 min.	1 to 5 f D.
" Bicarb.	8 to 60 gr.	50 cts. to 4 Gm.
" Bichrom.	$\frac{1}{12}$ to $\frac{1}{4}$ gr.	5 to 15 mills.
" Bitartr.	1 to 2 dr.	4 to 8 Gm.
" Bromid.	8 to 60 gr.	50 cts. to 4 Gm.
" Carb.	8 to 30 gr.	50 cts. to 2 Gm.
" Chloras	8 to 30 gr.	50 cts. to 2 Gm.
" Chlorid.	8 to 30 gr.	50 cts. to 2 Gm.
" Citr.	15 to 30 gr.	1 to 2 Gm.
" Cyanid.	$\frac{1}{16}$ to $\frac{1}{8}$ gr.	4 to 8 mills.
" et Sod. Tartr.	$\frac{1}{2}$ to 1 oz.	15 to 30 Gm.
" Ferrocyan.	5 to 15 gr.	30 to 100 cents.
" Hypophosphis	5 to 30 gr.	30 cts. to 2 Gm.
" Iodid.	2 to 15 gr.	10 cts. to 1 Gm.

Remedies.	Dose in Apothecaries' Weights and Measures.	Dose in Metric Terms.
Potass Iodohydrarg.	⅛ to ¾ gr.	8 to 50 mills.
" Nitras	8 to 15 gr.	50 to 100 cents.
" Permang.	⅛ to 1 gr.	1 to 6 cents.
" Salicyl.	5 to 15 gr.	30 to 100 cents.
" Sulphas	3 to 5 dr.	12 to 20 Gm.
" Sulphis	15 to 60 gr.	1 to 4 Gm.
" Sulphuret.	1 to 10 gr.	6 to 60 cents.
" Tartras { lax.	1 to 2 dr.	4 to 8 Gm.
{ purg.	2 to 8 dr.	8 to 30 Gm.
Potentilla	30 to 60 gr.	2 to 4 Gm.
Prinos	30 to 60 gr.	2 to 4 Gm.
Propylamina (and salts)	2 to 15 gr.	10 to 100 cents.
Prunin (eclectic)	2 to 3 gr.	10 to 20 cents.
Prun. Virg.	15 to 60 gr.	1 to 4 Gm.
" " Inf.	2 to 3 fl. oz.	50 to 100 f Gm.
" " Syr.	1 to 4 fl. dr.	5 to 15 f Gm.
Ptelea	8 to 30 gr.	50 cts. to 2 Gm.
Ptelein (eclectic)	1 to 3 gr.	5 to 20 cents.
Pulegium	1 to 4 dr.	5 to 15 cents.
Pulmonaria	1 to 2 dr.	4 to 8 f Gm.
Pulsatilla .	1 to 5 gr.	5 to 30 cents.
Pulv. Aloë Canella	8 to 20 gr.	50 to 120 cents.
" Antimonialis	3 to 10 gr.	15 to 60 cents.
" Aromat.	8 to 30 gr.	50 cts. to 2 Gm.
" Arsen. Asiat.	1 to 3 gr.	5 to 20 cents.
" Catechu Co.	15 to 30 gr.	1 to 2 Gm.
" Cinchona Co.	1 to 2 dr.	4 to 8 Gm.
" Cinnam. Co.	8 to 30 gr.	50 cts. to 2 Gm.
" Creta Arom.	½ to 1 dr.	2 to 4 Gm.
" " " Opium	8 to 30 gr.	50 cts. to 2 Gm.
" " Co.	8 to 30 gr.	50 cts. to 2 Gm.
" Elater. Co.	½ to 5 gr.	3 to 30 cents.
" Glycyrrh. Co.	30 to 60 gr.	2 to 4 Gm.
" Ipecac. Co.	5 to 15 gr.	30 to 100 cents.
" Jalapa Co.	30 to 60 gr.	2 to 4 Gm.
" Kino Co.	5 to 20 gr.	30 to 120 cents.

Remedies.	Dose in Apothecaries' Weights and Measures.	Dose in Metric Terms.
Pulv. Morph. Sulph. Co.	8 to 15 gr.	50 to 100 cents.
" Opium Co.	5 to 15 gr.	30 to 100 cents.
" Rheum Co.	30 to 60 gr.	2 to 4 Gm.
" Scammon. Co.	8 to 20 gr.	50 to 120 cents.
Pycnanthem	1 to 2 dr.	4 to 8 Gm.
Pyrethrum	30 to 60 gr.	2 to 4 Gm.
" Tinct.	2 to 8 fl. dr.	8 to 30 f Gm.
Quassia	30 to 60 gr.	2 to 4 Gm.
" Extr.	3 to 5 gr.	15 to 30 cents.
" Inf.	1 to 2 fl. oz.	30 to 60 f Gm.
" Tinct.	½ to 2 fl. dr.	2 to 8 f Gm.
Quebracho Bark	10 to 45 gr.	50 cts. to 3 Gm.
Quercus	30 to 60 gr.	2 to 4 Gm.
" Decoct.	1 to 2 fl. oz.	30 to 60 f Gm.
Quinetum	1 to 20 gr.	5 to 120 cents.
Quinina	1 to 20 gr.	5 to 120 cents.
Quinin. Arsen.	$\frac{1}{32}$ to $\frac{1}{10}$ gr.	2 to 6 mills.
Quin. Benzoas	1 to 20 gr.	5 to 120 cents.
" Bromid.	1 to 10 gr.	5 to 60 cents.
" Chlorid.	1 to 20 gr.	5 to 120 cents.
" Kinas	1 to 20 gr.	5 to 120 cents.
" Mur.	1 to 20 gr.	5 to 120 cents.
" Phosph.	1 to 20 gr.	5 to 120 cents.
" Salicyl.	1 to 15 gr.	5 to 100 cents.
" Sulph.	1 to 20 gr.	5 to 120 cents.
" Tinctura	1 to 2 fl. dr.	4 to 8 f Gm.
" Tannas	1 to 20 gr.	5 to 120 cents.
" Valer.	1 to 3 gr.	5 to 20 cents.
Quinidina	1 to 20 gr.	5 to 120 cents.
Quinid. Sulph.	1 to 20 gr.	5 to 120 cents.
Resina Copaiba	2 to 10 gr.	10 to 60 cents.
" Jalapa	2 to 5 gr.	10 to 30 cents.
" Leptandra	2 to 4 gr.	10 to 25 cents.
" Podoph.	⅛ to ½ gr.	8 to 30 mills.
" Scammon.	5 to 15 gr.	25 to 100 cents.
Rhamnus Baccæ	15 to 40 gr.	1 to 2.50 Gm.

Remedies.	Dose in Apothecaries' Weights and Measures.	Dose in Metric Terms.
Rhamnus Succus	½ to 1 fl. oz.	15 to 30 f Gm.
" Frangula	½ to 2 dr.	2 to 8 Gm.
" Pursh.	½ to 2 dr.	2 to 8 Gm.
" " Extr.	3 to 10 gr.	15 to 60 cents.
Rhein (eclectic)	1 to 4 gr.	5 to 25 cents.
Rheum	1 to 30 gr.	5 to 200 cents.
" Extr.	5 to 15 gr.	30 to 100 cents.
" Fluidextr.	5 to 40 min.	3 to 25 f D.
" Pil.	1 to 5 pills.	1 to 5 pills.
" Pulv. Co.	15 to 60 gr.	1 to 4 Gm.
" Syr.	1 to 4 fl. dr.	5 to 15 f Gm.
" " Arom.	1 to 2 fl. dr.	4 to 8 f Gm.
" Tinct.	1 to 8 fl. dr.	5 to 30 f Gm.
" " Arom.	15 to 60 min.	1 to 4 f Gm.
" " Dulc.	1 to 4 fl. dr.	5 to 15 f Gm.
" " Senna	½ to 2 fl. dr.	2 to 8 f Gm.
Rhus Glabra	1 to 2 dr.	4 to 8 Gm.
Rhusin (eclectic)	1 to 2 gr.	5 to 12 cents.
Rhus Toxicod.	1 to 5 gr.	5 to 30 cents.
" " Tinct.	1 to 15 min.	1 to 10 f D.
Ricinus Folia	1 to 4 dr.	5 to 15 Gm.
" Oleum	2 to 8 fl. dr.	8 to 30 f Gm.
Rosa Gall.	2 to 6 dr.	4 to 25 Gm.
" Confect.	1 to 2 dr.	4 to 8 Gm.
" Syrup	1 to 2 fl. dr.	4 to 8 f Gm.
Rottlera	1 to 2 dr.	4 to 8 Gm.
Rubia	30 to 60 min.	2 to 4 f Gm.
Rubus	30 to 60 min.	2 to 4 f Gm.
Rudbeckia	1 to 2 dr.	4 to 8 Gm.
Rumex	30 to 60 min.	2 to 4 f Gm.
Rumin (eclectic)	1 to 5 gr.	5 to 30 cents.
Ruta	15 to 30 gr.	1 to 2 Gm.
Sabadilla	5 to 30 gr.	25 cts. to 2 Gm.
Sabbatia	30 to 60 gr.	2 to 4 Gm.
Sabina	5 to 15 gr.	25 cts. to 1 Gm.
Salicinum	8 to 30 gr.	50 cts. to 2 Gm.

Remedies.	Dose in Apothecaries' Weights and Measures.	Dose in Metric Terms.
Salix	½ to 2 dr.	2 to 8 Gm.
Salvia	15 to 30 gr.	1 to 2 Gm.
" Inf.	1 to 2 fl. oz.	30 to 60 f Gm.
Sanguinaria ⎰ expect.	1 to 5 gr.	5 to 30 cents.
⎱ emet.	10 to 20 gr.	50 to 120 cents.
" Acet.	15 to 30 min.	1 to 2 f Gm.
" Syrup.	30 to 60 min.	2 to 4 f Gm.
" Tinct. ⎰ expect.	15 to 60 min.	1 to 4 f Gm.
⎱ emet.	3 to 4 fl. dr.	10 to 15 f Gm.
Sanguinarin (eclectic)	1 to 2 gr.	5 to 12 cents.
Santal. Alb.	1 to 2 dr.	4 to 8 Gm.
" Aetherol.	15 to 40 min.	10 to 25 f D.
Santonica	8 to 60 gr.	50 cts. to 2 Gm.
" Extr.	3 to 10 gr.	15 to 60 cents.
Santoninum	2 to 10 gr.	10 to 60 cents.
Santon. Sod.	3 to 10 gr.	15 to 60 cents.
Sapo	5 to 30 gr.	25 cts. to 2 Gm.
Sarracenia	15 to 30 min.	1 to 2 f Gm.
Sarsapar.	30 to 60 gr.	2 to 4 Gm.
" Decoct.	1 to 6 fl. oz.	30 to 200 f Gm.
" Syr.	1 to 4 fl. dr.	5 to 15 f Gm.
Sassafras	30 to 60 gr.	2 to 4 Gm.
Scammon.	4 to 15 gr.	20 cts. to 1 Gm.
" Mistura	½ to 2 fl. oz.	15 to 60 f Gm.
" Pulv. Co.	8 to 20 gr.	50 to 120 cents.
" Resin	5 to 15 gr.	25 to 100 cents.
Scilla	1 to 2 gr.	5 to 12 cents.
" Acet.	15 to 40 min.	10 to 25 f D.
" Syr.	½ to 1 fl. dr.	3 to 4 f Gm.
" " Co.	15 to 60 min.	1 to 4 f Gm.
Scilla Tinct.	10 to 20 min.	6 to 12 f D.
Scoparius	15 to 60 gr.	1 to 4 Gm.
" Decoct.	1 to 4 fl. oz.	30 to 120 f Gm.
Scutellaria	1 to 2 dr.	4 to 8 Gm.
Scutellarin (eclectic)	1 to 2 gr.	5 to 12 cents.
Senecin (eclectic)	1 to 3 gr.	5 to 20 cents.

Remedies.	Dose in Apothecaries' Weights and Measures.	Dose in Metric Terms.
Senecio	1 to 2 dr.	4 to 8 Gm.
Senega	15 to 30 gr.	1 to 2 Gm.
" Decoct.	1 to 2 fl dr.	4 to 8 f Gm.
" Extract.	1 to 3 gr.	5 to 20 cents.
" Syrup.	1 to 2 fl. dr.	4 to 8 f Gm.
" Tinct.	½ to 2 fl. dr.	2 to 8 f Gm.
Senna	8 to 30 gr.	50 cts. to 2 Gm.
" Conf.	1 to 2 dr.	1 to 8 Gm.
" Infus. Co.	1 to 2 fl. oz.	30 to 60 f Gm.
" Syrup.	1 to 2 fl. dr.	4 to 8 f Gm.
" Tinct.	2 to 8 fl. dr.	8 to 30 f Gm.
Serpent.	8 to 15 gr.	50 to 100 cents.
" Infus.	1 to 2 fl. oz.	30 to 60 f Gm.
" Tinct.	½ to 2 fl. dr.	15 to 60 f Gm.
Serum Lactis	2 to 6 fl. oz.	60 to 200 f Gm.
Silphium	15 to 40 gr.	1 to 2.50 Gm.
Simaruba	15 to 60 gr.	1 to 4 Gm.
Sinapis Alba	½ to 2 dr.	2 to 8 Gm.
Smilacin (eclectic)	2 to 5 gr.	10 to 30 cents.
Sod. Acct.	15 to 60 gr.	1 to 4 Gm.
" Arsen.	$\frac{1}{20}$ to $\frac{1}{8}$ gr.	3 to 8 mills.
" Benzoas	10 to 20 gr.	50 to 120 cents.
" Biboras	10 to 30 gr.	50 cts. to 2 Gm.
" Bicarb.	8 to 30 gr.	50 cts. to 2 Gm.
" Bisulphis	8 to 20 gr.	50 to 120 cents.
" Boras	8 to 30 gr.	50 cts. to 2 Gm.
" Brom.	5 to 60 gr.	25 cts. to 4 Gm.
" Carb.	8 to 30 gr.	50 cts. to 2 Gm.
" " Exsicc.	5 to 20 gr.	25 to 120 cents.
" Chloras	5 to 30 gr.	25 cts. to 2 Gm.
" Chlorid.	15 gr. to 1 oz.	1 to 30 Gm.
" Choleas	5 to 15 gr.	25 cts. to 1 Gm.
" Hypophosph.	8 to 30 gr.	50 cts. to 2 Gm.
" Hyposulphis	15 to 30 gr.	1 to 2 Gm.
" Iodid.	5 to 10 gr.	25 to 60 cents.
" Nitr.	8 to 60 gr.	50 cts. to 4 Gm.

Remedies.	Dose in Apothecaries' Weights and Measures.	Dose in Metric Terms.
Sod. Phosph.	2 to 8 gr.	10 to 50 cents.
" Salicyl.	5 to 30 gr.	25 cts. to 2 Gm.
" Santonas	3 to 10 gr.	15 to 60 cents.
" Sulphas	1 to 8 dr.	4 to 30 Gm.
" Sulphis	8 to 30 gr.	50 cts. to 2 Gm.
" Sulpho-Carbol.	5 to 20 gr.	25 to 120 cents.
" Sulpho-Vinas	1 to 4 dr.	5 to 15 Gm.
" Sulphuret.	¼ to 1 gr.	1 to 6 cer.ts
Solidago	½ to 1 dr.	2 to 4 Gm.
Spigelia	15 gr. to 2 dr.	1 to 8 Gm.
" Infus.	1 to 8 fl. oz.	30 to 250 f Gm.
Spiræa	8 to 60 gr.	50 cts. to 4 Gm.
Spir. Aether Co.	30 to 60 min.	2 to 4 f Gm.
" " Nitr.	½ to 2 fl. dr.	2 to 8 f Gm.
" Ammonia	8 to 30 min.	5 to 20 f D.
Spir. Ammon. Arom.	15 to 60 min.	1 to 4 f Gm.
" " Fœtid.	30 to 60 min.	2 to 4 f Gm.
" Anisum	30 to 60 min.	2 to 4 f Gm.
" Armorac. Co.	1 to 3 fl. dr.	4 to 12 f Gm.
" Cajuput.	30 to 60 min.	2 to 4 f Gm.
" Camph.	8 to 30 min.	5 to 20 f D.
" Chlorof.	15 to 60 min.	1 to 4 f Gm.
" Cinnam.	30 to 60 min.	2 to 4 f Gm.
" Frument.	2 to 8 fl. dr.	8 to 30 f Gm.
" Junip.	30 to 60 min.	2 to 4 f Gm.
" " Co.	30 to 60 min.	2 to 4 f Gm.
" Lavand.	30 to 60 min.	2 to 4 f Gm.
" " Co.	30 to 50 min.	2 to 4 f Gm.
" Limonis	30 to 60 min.	2 to 4 f Gm.
" Mentha Pip.	30 to 60 min.	2 to 4 f Gm.
" " Vir.	30 to 60 min.	2 to 4 f Gm.
" Myrist.	30 to 60 min.	2 to 4 f Gm.
" Rosmarini	8 to 30 min.	5 to 20 f D.
" Vin. Gall.	2 to 8 fl. dr.	8 to 30 f Gm.
Statice	30 to 60 gr.	2 to 4 Gm.
Stillingia	15 to 40 gr.	1 to 2.50 Gm.

Remedies.	Dose in Apothecaries' Weights and Measures.	Dose in Metric Terms.
Stillingia Fluid extr. Co.	3 to 15 min.	2 to 10 f D.
" Syr. Co.	1 to 2 fl. dr.	4 to 10 f Gm.
Stillingin (eclectic)	1 to 3 gr.	5 to 20 cents.
Stramon. Fol.	2 to 3 gr.	10 to 20 cents.
" " Extr.	¼ to 1 gr.	1 to 6 cents.
" " Tinct.	10 to 20 min.	6 to 12 f D.
" Sem.	1 to 2 gr.	5 to 12 cents.
" " Fluidextr.	1 to 3 min.	1 to 2 f D.
" " Tinct.	10 to 20 min.	6 to 12 f D.
Strychnina (and salts)	$\frac{1}{48}$ to $\frac{1}{12}$ gr.	1 to 5 mills.
Styrax	10 to 20 gr.	50 to 125 cents.
Succus Bellad.	5 to 15 min.	3 to 10 f D.
" Conium	30 to 60 min.	2 to 4 f Gm.
" Hyoscyam.	30 to 60 min.	2 to 4 f Gm.
" Junip.	1 to 4 fl. dr.	5 to 15 f Gm.
" Limonis	½ to 4 fl. oz.	15 to 120 f Gm.
" Sambucus	1 to 6 fl. dr.	4 to 25 f Gm.
" Taraxac.	2 to 4 fl. dr.	8 to 15 f Gm.
Sulphur Lot.	½ to 4 dr.	2 to 15 Gm.
" Præcip.	½ to 2 dr.	2 to 8 Gm.
" Sublim.	½ to 4 dr.	2 to 15 Gm.
Sumbul	15 to 60 gr.	1 to 4 Gm.
" Tinct.	10 to 30 min.	5 to 20 f D.
Symphyt.	1 to 2 dr.	4 to 8 Gm.
Syrup. Acacia	1 to 2 fl. dr.	4 to 8 Gm.
" Allium	1 to 4 fl. dr.	5 to 15 f Gm.
" Althaea	1 to 4 fl. dr.	5 to 15 f Gm.
" Amygd.	1 to 4 fl. dr.	5 to 15 f Gm.
" Aurant.	1 to 2 fl. dr.	4 to 8 f Gm.
" Calx	15 to 30 min.	10 to 20 f D.
" Ferr. Brom.	15 to 60 min.	1 to 4 f Gm.
" " Iod.	15 to 60 min.	1 to 4 f Gm.
" Glycyrrh.	1 to 2 fl. dr.	4 to 8 f Gm.
" Hemidesmus	1 to 4 fl. dr.	5 to 15 f Gm.
" Hypophosphit.	1 to 2 fl. dr.	4 to 8 f Gm.
" Ipecac. { expect.	½ to 1 fl. dr.	2 to 4 f Gm.
Ipecac. { emet.	4 to 6 fl. dr.	15 to 200 f Gm.

Remedies.	Dose in Apothecaries' Weights and Measures.	Dose in Metric Terms.
Syrup. Kramer.	½ to 4 fl. dr.	2 to 15 f Gm.
" Lactucar.	2 to 3 fl. dr.	8 to 12 f Gm.
" Papaver	1 to 2 fl. dr.	4 to 8 f Gm.
" Prunus Virg.	1 to 4 fl. dr.	5 to 15 f Gm.
" Rhamnus Cath.	1 to 2 fl. dr.	4 to 8 f Gm.
" Rheum	1 to 4 fl. dr.	5 to 15 f Gm.
" " Arom.	1 to 2 fl. dr.	4 to 8 f Gm.
" Rhœas	1 to 2 fl. dr.	4 to 8 f Gm.
" Rosa Gall.	1 to 2 fl. dr.	4 to 8 f Gm.
" Rubus	1 to 2 fl. dr.	4 to 8 f Gm.
" Sarsap. Co.	1 to 4 fl. dr.	5 to 15 f Gm.
" Scilla	½ to 1 fl. dr.	2 to 4 f Gm.
" " Co.	15 to 60 min.	10 to 40 f D.
" Senega	1 to 2 fl. dr.	4 to 8 f Gm.
" Senna	1 to 2 fl. dr.	4 to 8 fGm.
" Stillingia Co.	1 to 2 fl. dr.	4 to 8 f Gm.
" Tolut.	1 to 2 fl. dr.	4 to 8 f Gm.
" Zingiber.	1 to 4 fl. dr.	5 to 15 f Gm.
Tabacum Inf. (in enema)	4 to 8 fl. oz.	15 to 30 f Gm.
" Tinct.	8 to 40 min.	5 to 25 f D.
" Vin.	8 to 40 min.	5 to 25 f D.
Tamarindus	1 to 2 dr.	4 to 8 Gm.
Tanacetum	30 to 60 gr.	2 to 4 Gm.
" Inf.	4 to 8 fl. dr.	15 to 30 Gm.
Taraxac.	½ to 2 dr.	2 to 8 Gm.
" Extr.	15 to 30 gr.	1 to 2 Gm.
Taxus Baccata	1 to 5 gr.	5 to 30 cents.
Terebinthina	15 to 60 gr.	1 to 4 Gm.
" Chian.	15 to 60 gr.	1 to 4 Gm.
Testa Præpar.	10 to 40 gr.	50 to 250 cents.
Thea	1 to 2 dr.	4 to 8 Gm.
Theina	1 to 5 gr.	5 to 30 cents.
Thebaina	⅛ to ¾ gr.	8 to 40 mills.
Theobromina	1 to 5 gr.	5 to 30 cents.
Thuja	30 to 60 gr.	2 to 4 Gm.
" Tinct.	30 to 60 min.	2 to 4 f Gm.

Remedies.	Dose in Apothecaries' Weights and Measures.	Dose in Metric Terms.
Thymol	½ to 2 gr.	3 to 12 cents.
Thymus	1 to 4 dr.	5 to 15 Gm.
Tiglium Oleum	⅙ to 2 gr.	1 to 12 cents.
Tilia	1 to 3 dr.	4 to 12 Gm.
Tinct. Aconit. Fol.	5 to 10 min.	3 to 6 f D.
" " Rad.	1 to 5 min.	1 to 3 f D.
" " " Flem.	1 to 3 min.	1 to 2 f D.
" Aloë	1 to 2 fl. dr.	4 to 8 f Gm.
" " Myrrha	1 to 2 fl. dr.	4 to 8 f Gm.
" Arnica	10 to 60 min.	6 to 40 f D.
" Asafœt.	30 to 60 min.	2 to 4 f Gm.
" Aurant.	1 to 2 fl. dr.	4 to 8 f Gm.
" Bellad. Fol.	5 to 15 min.	3 to 10 f D.
" " Rad.	3 to 10 min.	2 to 6 f D.
" Benzoin	15 to 30 min.	1 to 2 f Gm.
" " Co.	½ to 2 fl. dr.	2 to 8 f Gm.
" Calamus	1 to 4 fl. dr.	5 to 15 f Gm.
" Calumba	1 to 4 fl. dr.	5 to 15 f Gm.
" Cannab. Ind.	5 to 15 min.	3 to 10 f D.
" Cantharid.	5 to 15 min.	3 to 10 f D.
" Capsic.	10 to 30 min.	6 to 20 f D.
" Cardam.	½ to 2 fl. dr.	2 to 8 f Gm.
" " Co.	½ to 2 fl. dr.	2 to 8 f Gm.
" Cascarilla	½ to 2 fl. dr.	2 to 8 f Gm.
" Castor	½ to 2 fl. dr.	2 to 8 f Gm.
" Catechu	½ to 2 fl. dr.	2 to 8 f Gm.
" Chinoidin	1 to 2 fl. dr.	4 to 8 f Gm.
" Chirata	15 to 60 min.	1 to 4 f Gm.
" Cimicif.	½ to 1 fl. dr.	2 to 4 f Gm.
" Cinch.	½ to 2 fl. dr.	2 to 8 f Gm.
" " Co.	½ to · 2 fl. dr.	2 to 8 f Gm.
" Cinnam.	1 to 2 fl. dr.	4 to 8 f Gm.
" Coca	½ to 2 fl. oz.	15 to 60 f Gm.
" Coccus	15 to 60 min.	1 to 4 f Gm.
" Colch. Rad.	5 to 20 min.	3 to 12 f D.
" " Sem.	10 to 30 min.	6 to 20 f D.

Remedies.	Dose in Apothecaries' Weights and Measures.	Dose in Metric Terms.
Tinct. Conium Fol.	½ to 1 fl. dr.	2 to 4 f Gm.
" " Fruct.	15 to 45 min.	1 to 3 f Gm.
" Coptis	30 to 60 min.	2 to 4 f Gm.
" Crocus	1 to 2 fl. dr.	4 to 8 f Gm.
" Cubeba	½ to 2 fl. dr.	2 to 8 f Gm.
" Damiana	1 to 3 fl. dr.	4 to 12 f Gm.
" Delphin.	5 to 15 min.	3 to 10 f D.
" Digital.	10 to 20 min.	6 to 12 f D.
" Erythroxylon	½ to 2 fl. oz.	30 to 60 f Gm.
" Eucalypt.	1 to 3 fl. dr.	4 to 12 f Gm.
" Ergota	10 to 60 min.	6 to 10 f D.
" Ferr. Acet.	10 to 30 min.	6 to 20 f D.
" " Chlor.	10 to 30 min.	6 to 20 f D.
" Gallæ	½ to 2 fl. dr.	2 to 8 f Gm.
" Gelsem.	8 to 30 min.	5 to 20 f D.
" Gent. Co.	½ to 2 fl. dr.	2 to 8 f Gm.
" Guaiacum	30 to 60 min.	2 to 4 f Gm.
" " Ammon.	30 to 60 min.	2 to 4 f Gm.
" Helleb.	30 to 60 min.	2 to 4 f Gm.
" Humulus	1 to 3 fl. dr.	4 to 12 f Gm.
" Hyoscyamus Fol.	15 to 60 min.	1 to 4 f Gm.
" " Sem.	10 to 40 min.	6 to 25 f D.
" Iodum	5 to 15 min.	3 to 10 f D.
" Iod. Comp.	5 to 20 min.	3 to 12 f D.
" Ipecac. { expect.	5 to 30 min.	3 to 20 f D.
{ emet.	2 to 6 fl. dr.	8 to 25 f Gm.
" Jalap.	½ to 2 fl. dr.	2 to 8 f Gm.
" Kino	½ to 2 fl. dr.	2 to 8 f Gm.
" Kramer.	½ to 2 fl. dr.	2 to 8 f Gm.
" Lobel.	15 to 60 min.	1 to 4 f Gm.
" " Aether.	15 to 60 min.	1 to 4 f Gm.
" Lupulin	½ to 2 fl. dr.	2 to 8 f Gm.
" Myrrha	30 to 60 min.	2 to 4 f Gm.
" Nux Vom.	8 to 30 min.	5 to 20 f D.
" Opium	5 to 30 min.	3 to 20 f D.
" " Acet.	5 to 20 min.	3 to 12 f D.

Remedies.	Dose in Apothecaries' Weights and Measures.	Dose in Metric Terms.
Tinct. Opium Ammon.	½ to 1 fl. dr.	2 to 4 f Gm.
" " Camph.	½ to 1½ fl. dr.	2 to 6 f Gm.
" " Deodor.	5 to 30 min.	3 to 20 f D.
" Phosphorus	8 to 40 min.	5 to 25 f D.
" Pilocarpus	½ to 2 fl. dr.	2 to 8 f Gm.
" Quassia	½ to 2 fl. dr.	2 to 8 f Gm.
" Quinina	1 to 2 fl. dr.	4 to 8 f Gm.
" " Ammon.	1 to 2 fl. dr.	4 to 8 f Gm.
" Rheum	1 to 8 fl. dr.	5 to 30 f Gm.
" " Arom.	15 to 60 min.	1 to 4 f Gm.
" " Dulc.	1 to 4 fl. dr.	5 to 15 f Gm.
" " Senna	½ to 2 fl. dr.	2 to 8 f Gm.
" Sanguin. { expect.	15 to 60 min.	1 to 4 f Gm.
{ emet.	3 to 4 fl. dr.	10 to 15 f Gm.
" Scilla.	10 to 20 min.	6 to 12 f D.
" Senega	½ to 2 fl. dr.	2 to 8 f Gm.
" Senna	2 to 8 fl. dr.	8 to 30 f Gm.
" Serpent.	½ to 2 fl. dr.	2 to 8 f Gm.
" Stramon. Fol.	10 to 20 min.	6 to 12 f D.
" " Sem.	5 to 15 min.	3 to 10 f D.
" Sumbul	8 to 30 min.	5 to 20 f D.
" Tabacum	8 to 40 min.	5 to 25 f D.
" Thuja	30 to 60 min.	2 to 4 f Gm.
" Tolut.	15 to 30 min.	1 to 2 f Gm.
" Toxicodendron	1 to 16 min.	1 to 10 f D.
" Valer.	½ to 2 fl. dr.	2 to 8 f Gm.
" " Ammon.	30 to 60 min.	2 to 4 f Gm.
" Vanilla	½ to 2 fl. dr.	2 to 8 f Gm.
" Veratr. Vir.	3 to 10 min.	2 to 6 f D.
" Zingiber	15 to 60 min.	1 to 4 f Gm.
Tolu Syr.	1 to 2 fl. dr.	4 to 8 f Gm.
" Tinct.	15 to 30 min.	1 to 2 f Gm.
Tormentilla	15 to 30 gr.	1 to 2 Gm.
Toxicodendron	1 to 5 gr.	5 to 30 cents.
" Tinct.	1 to 15 min.	1 to 10 f D.
Trifolium	15 to 60 gr.	1 to 4 Gm.

Remedies.	Dose in Apothecaries' Weights and Measures.	Dose in Metric Terms.
Trilliin (eclectic)	2 to 4 gr.	10 to 25 cents.
Trillium	30 to 60 gr.	2 to 4 Gm.
Trimethylamina	3 to 15 gr.	20 to 100 cents.
Triosteum	15 to 30 gr.	1 to 2 Gm.
Tritic. Rep.	2 to 8 dr.	8 to 30 Gm.
Turpeth Mineral emet.	2 to 5 gr.	10 to 30 cents.
Tussilago	1 to 3 dr.	4 to 12 Gm.
Urtica	5 to 20 gr.	30 to 120 cents.
Ustilago	10 to 60 gr.	60 cts. to 4 Gm.
Uva Ursi	15 to 60 gr.	1 to 4 Gm.
" Decoct.	1 to 2 fl. oz.	30 to 60 f Gm.
Vaccin. Crassifol.	15 to 60 gr.	1 to 4 Gm.
Valer.	10 to 30 gr.	60 cts. to 2 Gm.
" Extr.	10 to 30 gr.	60 cts. to 2 Gm.
" Inf.	1 to 2 fl. oz	30 to 60 f Gm.
" Tinct.	½ to 2 fl. dr.	2 to 8 f Gm.
" " Ammon.	30 to 60 min.	2 to 4 f Gm.
Vanilla	2 to 8 gr.	10 to 50 cents.
" Tinct.	½ to 2 fl. dr.	2 to 8 f Gm.
Veratrin (eclectic)	⅛ to ½ gr.	8 to 30 mills.
Veratrina (alkaloid)	$\frac{1}{48}$ to ⅙ gr.	1 to 10 mills.
Veratr. Alb.	3 to 6 gr.	15 to 30 cents.
" Vir.	3 to 6 gr.	15 to 30 cents.
" " Tinct.	3 to 10 min.	2 to 6 f D.
Verbascum	1 to 4 dr.	5 to 15 Gm.
Viburnin (eclectic)	1 to 3 gr.	5 to 20 cents.
Viburn. Opul.	1 to 2 dr.	4 to 8 Gm.
" Prunifol.	30 to 60 gr.	2 to 4 Gm.
" " Extr.	3 to 8 gr.	20 to 40 cents.
Vin. Aloë	1 to 2 fl. dr.	4 to 8 f Gm.
" Antim.	15 to 60 min.	1 to 4 f Gm.
" Arom.	1 to 4 fl. dr.	5 to 15 f Gm.
" Aurant. Amar.	30 to 120 min.	2 to 8 f Gm.
" Colch. Rad.	8 to 30 min.	5 to 20 f D.
" " Sem.	15 to 45 min.	1 to 3 f Gm.
" Ergota	1 to 3 fl. dr.	4 to 12 f Gm.

Remedies.	Dose in Apothecaries' Weights and Measures.	Dose in Metric Terms.
Vin. Ferrat. Amar.	1 to 2 fl. dr.	4 to 8 f Gm.
" Ferr. Citr.	1 to 2 fl. dr.	4 to 8 f Gm.
" Glycyrrh. Opium	15 to 45 min.	1 to 3 f Gm.
" Ipecac. { expect.	5 to 30 min.	3 to 20 f D.
{ emet.	3 to 6 fl. dr.	10 to 25 f Gm.
" Opium	8 to 45 min.	5 to 30 f D.
" " Croc.	2 to 8 min.	1 to 5 f D.
" Pix Liqu.	1 to 4 fl. dr.	5 to 15 f Gm.
" Port.	2 to 15 fl. dr.	10 to 60 f Gm.
" Quin.	30 to 60 min.	2 to 4 f Gm.
" Rheum	1 to 2 fl. dr.	4 to 8 f Gm.
" " Aqua	4 to 8 fl. dr.	15 to 30 f Gm
" Tabacum	8 to 40 min.	5 to 25 f D.
" Xericum	2 to 16 fl. dr.	10 to 60 f Gm.
Viola Ped.	30 to 60 gr.	2 to 4 Gm.
" Tricol.	1 to 4 dr.	5 to 15 Gm.
Viscum Alb.	½ to 1 dr.	2 to 4 Gm.
Xanthium	15 to 30 gr.	1 to 2 Gm.
Xanthorrhiza	30 to 60 gr.	2 to 4 Gm.
Xanthoxylin (eclectic)	1 to 2 gr.	5 to 12 cents.
Xanthoxylum	5 to 30 gr.	25 cts. to 2 Gm.
Xylolum	5 to 15 min.	3 to 10 f D.
Zinc. Acet.	1 to 2 gr.	5 to 12 cents.
" Brom.	½ to 2 gr.	3 to 12 cents.
" Carb.	2 to 10 gr.	10 to 60 cents.
" Chlorid.	½ to 2 gr.	3 to 12 cents.
" Iodid.	½ to 4 gr.	3 to 25 cents.
" Oxid.	2 to 10 gr.	10 to 60 cents.
" Phosphas	1 to 3 gr.	5 to 20 cents.
" Phosphid.	1/12 to ⅙ gr.	5 to 10 mills.
" Sulphas { Ton.	1 to 2 gr.	5 to 12 cents.
{ emet.	10 to 30 gr.	60 cts. to 2 Gm.
" Valer.	1 to 6 gr.	5 to 35 cents.
Zingiber	10 to 20 gr.	60 to 120 cents.
" Oleoresina	½ to 2 gr.	3 to 12 cents.
" Syr.	1 to 2 fl. dr.	4 to 8 f Gm.
" Tinct.	15 to 60 min.	1 to 4 f Gm.

INDEX.

16

BOOKS PUBLISHED

BY

PRESLEY BLAKISTON,

ON

CHEMISTRY, PHARMACY, Etc.

CHEMISTRY, INORGANIC AND ORGANIC.

With Experiments. By CHARLES L. BLOXAM, Professor
of Chemistry in King's College, London, and in the
Department for Artillery Studies, Woolwich. Fourth
edition. With nearly 300 Engravings. 8vo., $4.00.

" Professor Bloxam has given us a most excellent and useful practical treatise. His 666 pages [now 700] are crowded with facts and
experiments, nearly all well chosen, and many quite new, even to
scientific men. . . . . It is astonishing how much information he often
conveys in a few paragraphs. We might quote fifty instances of this."
—*Chemical News.*

By the same Author.

LABORATORY TEACHING; or, Progressive

Exercises in Practical Chemistry. Fourth edition.
With 83 Engravings. Crown 8vo., $1.75.

AN INTRODUCTION TO PHARMACEUTICAL

and Medical Chemistry. Part I—Theoretical and De-
scriptive. Part II—Practical and Analytical. Arranged
on the Principle of the course of Lectures on Chemistry as
delivered at, and the Instruction given in the Laboratories
of, the South London School of Pharmacy. By JOHN
MUTER, M.A., President of the Society of Public Analysis,
etc. A second edition, enlarged and rearranged. The two
Parts bound in one large octavo volume. Price, $6.00.

A PHARMACOPŒIA, INCLUDING THE OUT-

lines of Materia Medica and Therapeutics, for the use of
Practitioners and Students of Veterinary Medicine. By
RICHARD V. TUSON, F.C.S., Professor of Chemistry, Ma-
teria Medica, and Toxicology, at the Royal Veterinary
College. Third edition. Post 8vo, $2.50.

" Not only practitioners and students of veterinary medicine, but
chemists and druggists will find that this book supplies a want in
veterinary literature."—*Chemist and Druggist.*

THE POCKET FORMULARY AND SYNOPSIS

of the British and Foreign Pharmacopœias. Comprising
Standard and approved Formulæ for the Preparation and
Compounds employed in Medical Practice. By HENRY
BEASLEY. Tenth edition. 18mo, $2.25.

By the same Author.

THE DRUGGIST'S GENERAL RECEIPT BOOK.

Comprising a Copious Veterinary Formulary, Numerous
Recipes in Patent and Proprietary Medicines, Druggists'
Nostrums, etc.; Perfumery and Cosmetics; Beverages,
Dietetic Articles and Condiments; Trade Chemicals;
Scientific Processes; and an Appendix of Useful Tables.
Eighth edition. 18mo, $2.25.

Also,

THE BOOK OF PRESCRIPTIONS: Containing

3107 Prescriptions collected from the Practice of the
most eminent physicians and Surgeons, English and
Foreign. With an Index of Diseases and Remedies.
Fifth edition. 18mo, $2.25.

" Mr. Beasley's ' Pocket Formulary,' ' Druggist's Receipt Book,' and
' Book of Prescriptions,' form a compact library of reference admirably
suited for the dispensing desk."—*Chemist and Druggist.*

AN INTRODUCTION TO THE PRACTICE OF

Commercial Organic Analysis. By ALFRED H. ALLEN,
F.C.S., Lecturer on Chemistry at the Sheffield School of
Medicine. Vol. I.—Cyanogen Compounds, Alcohols
and their Derivatives, Phenols, Acids, etc. 8vo, $3.50.

THE ART OF PERFUMERY: The Methods of

Obtaining the Odors of Plants and Instruction for the
Manufacture of Perfumery, Dentifrices, Soap, etc., etc.
By G. W. SEPTIMUS PIESSE. Fourth edition, enlarged.
366 illustrations. 8vo. Cloth. Price, $5.50.

" An excellent book."—*Commercial Advertiser.*
" It is the best book on Perfumery yet published."—*Scientific Amer.*
" Exceedingly useful to druggists and perfumers."—*Jour. of Chem.*
" Is in the fullest sense, comprehensive."—*Medical Record.*